Plant Stress Physiology - Perspectives in Agriculture

Edited by Mirza Hasanuzzaman and Kamran Nahar

Published in London, United Kingdom

IntechOpen

Supporting open minds since 2005

Plant Stress Physiology - Perspectives in Agriculture
http://dx.doi.org/10.5772/intechopen.94821
Edited by Mirza Hasanuzzaman and Kamran Nahar

Contributors
Paola Andrea Palmeros-Suárez, Julio Armando Massange-Sánchez, Carla Vanessa Sánchez-Hernández, Rosalba Mireya Hernández-Herrera, Alexandru Dascaliuc, Mhemmed Gandour, Mirela Irina Cordea, Orsolya Borsai, Fatemeh Habibpourmehraban, Melekote Nagabhushan Arun, Shibara Shankara Hebbar, Bhanuprakash, Thulasiram Senthivel, Anil Kumar Nair, Guntupalli Padmavathi, Pratima Pandey, Aarti Singh, Sambangi Pratyusha, Parul Chowdhury, Ambreen Bano, Anmol Gupta, Neelam Pathak, Smita Rai, Swati Sharma, Touseef Fatima, Mirza Hasanuzzaman, Khursheda Parvin, Taufika Islam Anee, Abdul Awal Chowdhury Awal Chowdhury Masud, Farzana Nowroz

First published in London, United Kingdom, 2022 by IntechOpen
IntechOpen is the global imprint of INTECHOPEN LIMITED, registered in England and Wales, registration number: 11086078, 5 Princes Gate Court, London, SW7 2QJ, United Kingdom
Printed in Croatia

British Library Cataloguing-in-Publication Data
A catalogue record for this book is available from the British Library

Additional hard and PDF copies can be obtained from orders@intechopen.com

Plant Stress Physiology - Perspectives in Agriculture
Edited by Mirza Hasanuzzaman and Kamran Nahar
p. cm.

This title is part of the Physiology Book Series, Volume 11
Topic: Plant Physiology
Series Editor: Tomasz Brzozowski
Topic Editor: Jen-Tsung Chen

Print ISBN 978-1-83969-866-8
Online ISBN 978-1-83969-867-5
eBook (PDF) ISBN 978-1-83969-868-2
ISSN 2631-8261

For EU product safety concerns:
IN TECH d.o.o., Prolaz Marije Krucifikse Kozulić 3, 51000 Rijeka, Croatia, info@intechopen.com or visit our website at intechopen.com.

IntechOpen Book Series
Physiology
Volume 11

Aims and Scope of the Series

Modern physiology requires a comprehensive understanding of the integration of tissues and organs throughout the mammalian body, including the cooperation between structure and function at the cellular and molecular levels governed by gene and protein expression. While a daunting task, learning is facilitated by identifying common and effective signaling pathways mediated by a variety of factors employed by nature to preserve and sustain homeostatic life. As a leading example, the cellular interaction between intracellular concentration of Ca+2 increases, and changes in plasma membrane potential is integral for coordinating blood flow, governing the exocytosis of neurotransmitters, and modulating gene expression and cell effector secretory functions. Furthermore, in this manner, understanding the systemic interaction between the cardiovascular and nervous systems has become more important than ever as human populations' life prolongation, aging and mechanisms of cellular oxidative signaling are utilised for sustaining life. Altogether, physiological research enables our identification of distinct and precise points of transition from health to the development of multimorbidity throughout the inevitable aging disorders (e.g., diabetes, hypertension, chronic kidney disease, heart failure, peptic ulcer, inflammatory bowel disease, age-related macular degeneration, cancer). With consideration of all organ systems (e.g., brain, heart, lung, gut, skeletal and smooth muscle, liver, pancreas, kidney, eye) and the interactions thereof, this Physiology Series will address the goals of resolving (1) Aging physiology and chronic disease progression (2) Examination of key cellular pathways as they relate to calcium, oxidative stress, and electrical signaling, and (3) how changes in plasma membrane produced by lipid peroxidation products can affect aging physiology, covering new research in the area of cell, human, plant and animal physiology.

Meet the Series Editor

Prof. Dr. Thomas Brzozowski works as a professor of Human Physiology and is currently a Chairman at the Department of Physiology and is V-Dean of the Medical Faculty at Jagiellonian University Medical College, Cracow, Poland. His primary area of interest is physiology and pathophysiology of the gastrointestinal (GI) tract, with a major focus on the mechanism of GI mucosal defense, protection, and ulcer healing. He was a postdoctoral NIH fellow at the University of California and the Gastroenterology VA Medical Center, Irvine, Long Beach, CA, USA, and at the Gastroenterology Clinics Erlangen-Nuremberg and Munster in Germany. He has published 290 original articles in some of the most prestigious scientific journals and seven book chapters on the pathophysiology of the GI tract, gastroprotection, ulcer healing, drug therapy of peptic ulcers, hormonal regulation of the gut, and inflammatory bowel disease.

Meet the Volume Editors

Dr. Mirza Hasanuzzaman is a Professor of Agronomy at Sher-e-Bangla Agricultural University, Bangladesh. He received his Ph.D. in Plant Stress Physiology and Antioxidant Metabolism from Ehime University, Japan, with a scholarship from the Japanese Government (MEXT). Later, he completed his postdoctoral research at the Center of Molecular Biosciences, University of the Ryukyus, Japan, as a recipient of the Japan Society for the Promotion of Science (JSPS) postdoctoral fellowship. He was also the recipient of the Australian Government Endeavour Research Fellowship for postdoctoral research as an adjunct senior researcher at the University of Tasmania, Australia. Dr. Hasanuzzaman's current work is focused on the physiological and molecular mechanisms of environmental stress tolerance. Dr. Hasanuzzaman has published more than 150 articles in peer-reviewed journals. He has edited ten books and written more than forty book chapters on important aspects of plant physiology, plant stress tolerance, and crop production. According to Scopus, Dr. Hasanuzzaman's publications have received more than 10,500 citations with an h-index of 53. He has been named a Highly Cited Researcher by Clarivate. He is an editor and reviewer for more than fifty peer-reviewed international journals and was a recipient of the "Publons Peer Review Award" in 2017, 2018, and 2019. He has been honored by different authorities for his outstanding performance in various fields like research and education, and he has received the World Academy of Science Young Scientist Award (2014) and the University Grants Commission (UGC) Award 2018. He is a fellow of the Bangladesh Academy of Sciences (BAS) and the Royal Society of Biology.

Dr. Kamrun Nahar is a Professor of Agricultural Botany at Sher-e-Bangla Agricultural University, Bangladesh. She received her Ph.D. in Environmental Stress Physiology of Plants from the United Graduate School of Agricultural Sciences, Ehime University, Japan, with a scholarship from the Japanese Government (MEXT). Dr. Nahar has been involved in research with field crops emphasizing stress physiology since 2006. She has completed several research works and is currently working on a research project funded by Sher-eBangla Agricultural University Research System and the Ministry of Science and Technology, Bangladesh. She is also supervising MS students. Dr. Nahar has published more than 100 articles and book chapters related to plant physiology and environmental stresses. Her publications have received about 9,500 citations with an h-index of 51. She is involved in editorial activities and is a reviewer of international journals. She is an active member of about twenty professional societies. Dr. Nahar has attended numerous international conferences and presented twenty papers and posters at these conferences.

Contents

Preface

Plants face several types of environmental stresses in their natural habitat and due to their sessile nature, they cannot move to favorable environmental conditions to avoid these stresses. These environmental stresses can be biotic (insects, pathogens) or abiotic (salinity, drought, flooding, extreme temperatures), both of which can retard plant growth and development. Plants have an intrinsic mechanism to regulate these stresses by synthesizing natural products through different metabolic pathways. These natural products can be primary or secondary metabolites and their synthesis is strictly bound to genetic and epigenetic controlled pathways. Further transportation and regulation of these products within plants depends on the differential responses of biotic and abiotic stresses (Isah 2019). Abiotic stresses are considered the most devastating; they are responsible for reducing major crop yields by about 50% worldwide (Saini et al. 2018). Salinity, drought, and flooding are among the main abiotic stresses. The extent of crop yield reduction due to these stresses largely depends on the developmental stage of the plant and the duration and intensity of stress that the plant is undergoing. Furthermore, in nature, abiotic stresses occur in combination, as they have an interrelated role in different biochemical pathways and impinged cellular, metabolic, and physiological activities of plants. To enhance plants' survival and production efficiency, scientists have developed different approaches to increase the tolerance and resistance capacity of plants under abiotic stresses. Recent advances in using different biotechnological tools and techniques lead to the identification of genes and manipulation of any specific characters in plants. Protein and metabolite profiling using techniques of biotechnology helps in understanding the complex physiological mechanisms of plants. Thus, knowledge of drought-tolerant mechanisms and their relation to different plant traits helps in the identification of quantitative trait loci and genes that are linked with other traits. Through the use of genetic engineering approaches, scientists have intensively investigated the expression of stress-responsive genes in plants under abiotic stress that have a potential role in plant defense signaling pathways. In addition, exogenous applications of different chemicals and biostimulants are being used to induce stress tolerance in plants and improve plant growth and development by enhancing different physiological attributes of plants. These exogenous products include phytohormones, humic substances, basic nutrients, plant or algal extracts, microorganisms, and so on that have the potential to enhance abiotic stress tolerance capacity in plants and maintain sustainable plant production in climate-resilient agriculture.

This book includes ten chapters dealing with various aspects of plant stress responses and tolerance. These chapters discuss various physiological and molecular plant mechanisms.

We would like to give special thanks to the authors for their outstanding and timely work in producing such excellent chapters. We are also thankful to Author Service Manager Zrinka Tomicic at IntechOpen for her assistance throughout the preparation and publication of this book. We believe this book is useful for undergraduate and graduate students, teachers, and researchers in the field of plant physiology and crop science.

Dr. Mirza Hasanuzzaman
Professor,
Department of Agronomy,
Sher-e-Bangla Agricultural University,
Dhaka, Bangladesh

Dr. Kamran Nahar
Professor,
Department of Agricultural Botany,
Sher-e-Bangla Agricultural University,
Dhaka, Bangladesh

Chapter 1

The Biochemical Mechanisms of Salt Tolerance in Plants

Julio Armando Massange-Sánchez, Carla Vanessa Sánchez-Hernández, Rosalba Mireya Hernández-Herrera and Paola Andrea Palmeros-Suárez

Abstract

Salinity is one of the most severe environmental problems worldwide and affects plant growth, reproduction, and crop yields by inducing physiological and biochemical changes due to osmotic and ionic shifts in plant cells. One of the principal modifications caused by osmotic stress is the accumulation of reactive oxygen species (ROS), which cause membrane damage and alter proteins, DNA structures, and photosynthetic processes. In response, plants increase their arsenal of antioxidant compounds, such as ROS scavenging enzymes and nonenzymatic elements like ascorbate, glutathione, flavonoids, tocopherols, and carotenoids, and their rates of osmolyte synthesis to conserve ion homeostasis and manage salt stress. This chapter describes the principal biochemical mechanisms that are employed by plants to survive under salt-stress conditions, including the most recent research regarding plant tolerance, and suggests strategies to produce valuable crops that are able to deal with soil salinity.

Keywords: salinity, ROS, scavenging enzyme, antioxidant compound, osmolyte

1. Introduction

Salt stress in the form of soil salinity restricts plant growth and limits crop yields. Globally, soil salinity affects 6–10% of the total land surface (~ 800 million ha), 20–33% of which is used for agricultural purposes [1–3]. The damages caused by soil salinity are more notable in arid and semiarid regions where limited rainfall, high evapotranspiration rates, and extreme temperatures coupled with poor water and soil management practices exacerbate this problem [4, 5]. Despite the current agricultural problems due to soil salinity, world population growth has exerted substantial pressure for increased crop production to meet the global demand for food. Simultaneously, climate change has continued to markedly challenge the growth and production of agricultural crops due to variations in temperature, shifts in precipitation, reduced solar radiation, and increased evaporative demand.

Salinization can be the result of natural causes, such as flooding, wind erosion, and the redistribution of salts in shallow groundwater systems due to the weathering of minerals and salty rocks that release sodium chloride (NaCl) and other soluble salts like magnesium, calcium, sulfates (SO4), and carbonates (CO3) into the environment [6, 7]. Other sources of salinity include high evaporation rates in the tropics and the precipitation of ocean salt by wind and rain, which increase ion concentrations in soils [8]. However, the global increase in salinization is mainly due to anthropogenic activities, such as irrigation, land clearing, deforestation, agricultural intensification, the use of imbalanced fertilizers, or poor drainage systems [7]. Irrigation practices have also elevated groundwater levels, resulting in a subsequent increase in evaporation. Moreover, industrial wastewater and effluents enriched in salts can elevate the salinity levels of agricultural soils.

Salinity has drastic consequences in plants at a physiological level because it restricts the ability of plants to take up water. In plants under salt-stress conditions, the rates of transpiration and photosynthesis decrease while the opening and closing of stomata is generally quick; in addition, ion toxicity, membrane instability, and mineral limitation are observed along with the inhibition of both enzymes and metabolic pathways [9–11]. Additional problems emerge when excessive amounts of salts enter a plant and reach toxic levels, impairing normal germination, plant growth, lateral bud development, productivity, and senescence [12]. The overall results of salinity stress can be seen in impaired plant growth, physiological functions, and crop yields, although these depend on the severity of the stress, the time scale of the response, and on whether the stress was abruptly or gradually imposed [11].

Plant damage caused by salinity primarily takes place in two phases. First, increases in soil salt content cause an osmotic effect, which reduces the soil water potential and consequently plant water absorption. Second, the excessive uptake of ions, mainly Na+, Cl-, and SO2, interferes with various metabolic processes in plants, including photosynthesis, nitrogen assimilation, malate metabolism, and protein translation [13]. This nutrient imbalance reduces the amounts of calcium, magnesium, and potassium in the plant [14]. In addition, oxidative damage is generated as a result of an increase in ROS due to a reduction in the rate of photosynthesis. The combination of these factors negatively affects plant functions, including those related to metabolism, and may damage plant structures, which can ultimately lead to plant death [15, 16].

The ability of crops to grow and reproduce in saline soil is dependent on the developmental stage of the plant and greatly differs between species, which is mainly due to the variability associated with limiting salt uptake from the soil and effectively compartmentalizing it at the cellular level [17]. To deal with the adverse effects of salinity, plants trigger different physiological and biochemical mechanisms that allow them to survive and grow, including salt exclusion, the control of ion uptake and translocation, ion compartmentalization within different cells and tissues, nutrient ion transport, the synthesis of compatible solutes and osmoprotectants, morphological and anatomical modifications, membrane and hormone changes, and antioxidative metabolism responses based on the production of antioxidant enzymes and compounds [8, 13, 18]. This chapter is focused on presenting the principal determinants of salt-stress tolerance in plants, which include the antioxidant defense system, enzymatic and nonenzymatic compounds, and the most important metabolites involved in osmotic adjustments.

2. Antioxidant defense mechanisms

The antioxidant defense system is a vital mechanism by which plants deal with oxidative stress under saline conditions [19–21]. Salinization reduces the rate of photosynthesis, which alters the electron flow from central transport chains to oxygen-reduction pathways, leading to an overproduction of ROS, such as superoxide radicals ($O2\bullet-$), singlet oxygen ($O2$), hydrogen peroxide ($H2O2$), hydroxyl radicals ($OH\bullet$), and alkoxy radicals ($RO\bullet$). ROS are generated in the cytosol, chloroplasts, mitochondria, and apoplastic spaces of cells [22–25]. Although ROS are formed in biological systems as by-products of respiration and photosynthesis, in low concentrations, they act as signal transduction molecules that are involved in mediating programmed cell death, development, and responses to pathogen infections and environmental stressors [23, 26]. However, an excess of ROS produces phytotoxic reactions, biomolecule oxidation, cell membrane damage, protein degradation, enzyme inhibition, and DNA mutations [27–29]. Therefore, fluctuations in the properties and functions of ROS will eventually lead to metabolic and physiological problems in plants.

The overproduction of ROS imbalances homeostasis at cellular and subcellular levels and may ultimately lead to cellular death [30]. The intensity of the oxidative stress depends on the types of ROS produced, their concentrations, the sites where they are released, interactions with other cellular molecules, and the developmental stage and potential of the cell [31]. To avoid damage caused by the overproduction of ROS, plants employ an antioxidant defense mechanism that contains several enzymatic and nonenzymatic constituents that mitigate potential negative effects by converting ROS into less toxic molecules [32]. Given that increases in antioxidant levels usually improve plant tolerance, an adjustable control system that balances ROS production and scavenging via antioxidant enzyme and non-enzyme activities is essential for limiting toxicity levels in plants under adverse conditions [33, 34].

Plants rely on enzymes that are specifically involved in ROS detoxification, namely those that act to lower ROS levels or avoid oxidative stress, including superoxide dismutase (SOD), catalase (CAT), ascorbate peroxidase (APX), and other enzymes involved in the AsA-GSH cycle, such as glutathione reductase (GR), monodehydroascorbate reductase (MDHAR), and dehydroascorbate reductase (DHAR). These antioxidant enzymes are located in different sites within plant cells and work together to detoxify ROS [35]. In addition, low molecular mass antioxidant compounds like ascorbate, glutathione, flavonoids, tocopherols, and carotenoids are also crucial for ROS homeostasis in plants [36–38].

2.1 Enzymatic antioxidants

2.1.1 Superoxide dismutase

Superoxide dismutase, which shows great variation across different plant species, is a metalloenzyme that plays an important role in combating oxidative stress in all subcellular organelles sensitive to ROS (i.e., chloroplasts, mitochondria, peroxisomes). This metalloenzyme catalyzes the dismutation of $O2\bullet -$ into $O2$ and $H2O2$. A particularly interesting property of SODs is that they can only react with $O2\bullet-$) through diffusion and electrostatic guidance at limited rates, yet they are highly effective at removing $O2\bullet-$ [39]. SOD enzymes can be classified into three types according to their metal cofactors: copper/zinc (Cu/Zn-SOD; found in the cytosol or in plastids), manganese (Mn-SOD; found in mitochondria and peroxisomes), or iron (Fe-SOD; found in chloroplasts). Of note, it is only plants that have all three SOD types [39]. Superoxide dismutases are considered the first line of defense against ROS because an increase in SOD activity has been frequently linked to improved plant tolerance in the presence of environmental stress [40]. For example, the overproduction of SOD has been found to enhance salt-stress tolerance in sweet potatoes [40]. Moreover, SOD activity has been put forth as an indirect means to screen for plants that are salinity resistant.

2.1.2 Catalase

Catalase was the first enzyme to be identified with antioxidant capabilities and is mainly found in peroxisomes and glyoxysomes. The function of CAT is to convert H2O2 into H20 and O2 [41]. H2O2 has been found to be produced in cell organelles during photorespiratory oxidation, the β-oxidation of fatty acids, purine catabolism, and because of the activities of enzymes, such as xanthine oxidase and SOD [42]. CAT plays a dynamic role in eliminating H2O2 by degrading it in an energetically efficient manner. For example, one CAT molecule can transform about 6 million H2O2 molecules into H2O and O2 in roughly one minute [43, 44].

Whereas animals contain a single CAT gene, plants have a multigene family that includes multiple isoenzymes (depending on the species) whose expression is regulated according to how they are distributed within tissues or organs and the environmental conditions present [44]. In plants, there are three classes of catalases that are determined based on location: class I (photosynthetic tissues), class II (vascular tissues), and class III (seeds and young seedlings) [45]. A concomitant increase in CAT activity has been observed as part of the antioxidant defense system in plants. This increase may be a manifestation of the adaptive responses of plants to abiotic stress. It may also be assumed that without any concomitant increase in CAT activity, plant growth is likely to be severely limited. Therefore, different environmental stressors constitute major drivers that either enhance or restrict CAT activity, depending on the intensity, duration, and type of stress [46].

2.1.3 Enzymes involved in the AsA-GSH cycle

The Asa-GSH cycle, also known as the Halliwell–Asada cycle, is a series of coupled redox reactions involving four enzymes: APX, MDHAR, DHAR, and GR [47]. This cycle plays a crucial role in the antioxidant protection system in the presence of H2O2, which is generated in different cellular compartments like chloroplasts, mitochondria, peroxisomes, and the cytosol and apoplast [48, 49]. H2O2 is scavenged by APX via the oxidation of ascorbate. This enzyme comprises a family of five APX isoforms that are located in several organelles and cellular compartments, including glyoxysomes (gmAPX), thylakoids (tAPX), the cytosol (cAPX), mitochondria (mAPX), and soluble chloroplast stroma (sAPX) [46]. APX uses two molecules of ascorbate to reduce H2O2 to H2O and produces two molecules of monodehydroascorbate (MDHA) from the oxidation of ascorbate. MHDA is converted back to ascorbate by MDHAR, which is a flavin adenine dinucleotide (FAD) enzyme with high specificity for MHDA. Subsequently, MDHA is further rapidly converted to dehydroascorbate (DHA) by the monomeric thiol enzyme DHAR. DHAR uses reduced glutathione (GSH) as the reducing substrate, which is regenerated by GR from its oxidized form glutathione disulfide (GSSG), and NADPH molecules act as donors of reducing equivalents [29, 43]. Recent studies have shown that these enzymes become active and participate in the ascorbic acid-glutathione cycle under salt-stress conditions, indicating that their increased activity may improve plant tolerance [50–52]. **Figure 1** shows the functions of the antioxidant enzymes described above.

2.2 Nonenzymatic antioxidants

2.2.1 Ascorbic acid

Ascorbic acid (well known as Vitamin C) is the main antioxidant in plants and acts as a major redox buffer and cofactor for enzymes. The structure of ascorbic acid is like those of hexose sugars and is comprised of a conjugated structure

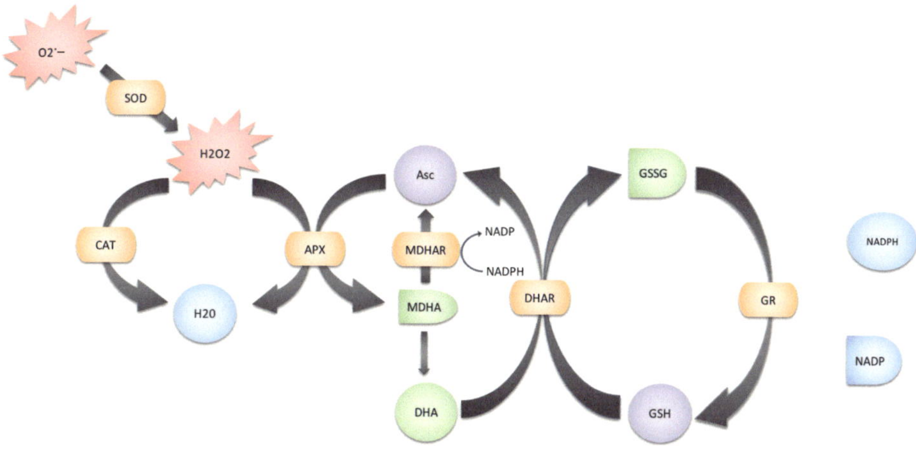

Figure 1.
The orange squares show ROS scavenging by the antioxidant enzymes superoxide dismutase (SOD) and catalase (CAT) and the enzymes involved in the Asa-GSH cycle, namely ascorbate peroxidase (APX), monodehydroascorbate (MDHAR), dehydroascorbate reductase (DHAR), and glutathione reductase (GR). The purple circles show the nonenzymatic antioxidant compounds glutathione and ascorbate that are described in section 2.2. The diagram was modified from [48].

composed of a five-carbon lactone ring, containing enediol groups on C2 and C3. Ascorbic acid is considered to be the most powerful ROS scavenger due to its ability to donate electrons to a wide range of electron receptors in several enzymatic and nonenzymatic reactions [53]. Ascorbic acid protects cellular membranes by directly scavenging O2• – and OH while acting as a cofactor for violaxanthin deepoxidase, sustaining the dissipation of excess excitation energy in chloroplasts [54]. In addition, ascorbic acid has the redox potential to interact with hydroxyl radicals,, superoxides, oxidized glutathione, and tocopherol radicals [53, 55].

Ascorbic acid is involved in many plant pathways related to photosynthesis, hormone biosynthesis, antioxidant regeneration, defense responses, signal transduction, flowering, cell division, and growth and is abundantly found in meristems and photosynthetic cells [56, 57]. Under normal physiological conditions, ascorbic acid largely remains in a reduced form in leaves and chloroplasts [55] and accumulates in photosynthetic organs, although it can be found in high concentrations in non-photosynthetic tissues with the amounts varying depending on the plant species, variety, tissue type, and growth environment [53, 58].

Multiple experiments have been developed to evaluate the roles of antioxidant compounds, such as glutathione and ascorbic acid, when they are applied to different plant tissues. For example, Aliniaeifard et al. [59] sprayed 2 mM AsA and 3 mM GSH on olive plants treated with 100 mM NaCl and reported that plant growth parameters significantly improved with the application of AsA when compared with those of the plants sprayed with GSH. In addition, the Na + and Cl– concentrations decreased in the olive plants sprayed with AsA while the K+ concentration and K/Na ratio increased. Aliniaeifard et al. concluded that the exogenous application of AsA is recommended for improving the tolerance of olive plants under saline conditions. Similarly, Sadak [60] applied AsA to flax cultivars irrigated with salt water and found that AsA allowed for an increase in antioxidant defense via the activation of ROS scavenging enzymes.

2.2.2 Glutathione

Together with ascorbic acid, glutathione is one of the most important nonenzymatic antioxidants. Glutathione is a thiol tripeptide composed of γ-glutamyl-cysteinylglycine and plays a central role in antioxidant defense by scavenging ROS, thereby maintaining redox homeostasis in plant tissues [61]. The chemical reactivity, relative stability, and high water solubility of the thiol group of glutathione makes it ideal for protecting plants against oxidative stress or stress due to heavy metals and exogenous or endogenous organic chemicals [62]. Glutathione is localized in all cell compartments, such as the chloroplasts, cytosol, vacuoles, mitochondria, and endoplasmic reticulum. In physiological processes, glutathione plays important roles in xenobiotic detoxification, metabolite conjugation, signal transduction, and stress-responsive gene expression [61, 63]. Glutathione also participates in tissue growth and development, cell death and senescence, and the enzymatic regulation of pathogen resistance [64].

Glutathione also appears to be important in controlling O2•−, •OH, and H2O2 levels in cells under stressful conditions given the observed changes in the ratio of its reduced (GSH) and oxidized (GSSG) forms, which provides information on the cellular redox balance [65, 66]. GSH is required for GRX-mediated dithiol and monothiol reduction mechanisms and is important for the activation of GPX, GST, and glyoxalase functions, particularly recycling GSH from GSSG in the presence of NADPH. Therefore, GSH plays a key role in ROS tolerance under oxidative stress in association with other ROS-metabolizing enzymes [57]. In addition, the primary reaction in plants when they are exposed to progressive salinity is the oxidation of GSH, indicating that this reaction is an early response to oxidative stress [33, 67].

Studies have demonstrated the benefits associated with the application of GSH in modulating salt-stress tolerance and positively influencing yield-contributing traits in plants like rice (*Oryza sativa* L.) [68], while the application of GSH has been found to increase the transcript levels and activities of genes and enzymes related to GSH synthesis and metabolism in tomato plants (*Solanum lycopersicum*) [69].

2.2.3 Flavonoids

Flavonoids are largely responsible for the pigmentation of seeds, flowers, stems, roots, and fruits. Flavonoids are aromatic compounds that are derived from chalcone, which is obtained from phenylalanine and malonyl-coenzyme A. According to their chemical structures, flavonoids can be classified into flavones, flavanones, flavonols, flavans, flavandiols, isoflavones, anthocyanins, and condensed tannins [70, 71].

Flavonoids act to prevent ROS increase through the inhibition of ROS-generating enzymes, including cyclooxygenase, lipoxygenase, monooxygenase, and xanthine oxidase. The ability to chelate metals is essential for impeding the regeneration of radical ROS molecules, such as superoxide, hydroxyl radicals, and hydrogen peroxide [72, 73]. Given that they contain functional hydroxyl groups, flavonoids constitute a secondary antioxidant system, as they support the function of other ROS scavenging systems when the activities of antioxidant enzymes are reduced. In addition, flavonoids serve as antioxidant barriers that protect cellular components against oxidizing pollutants like ozone (O3) and sulfur dioxide (SO2). Due to their elevated abilities to donate electrons or hydrogen atoms, the antioxidant capacities of flavonoids have been found to be many times higher than those of either ascorbic acid or a-tocopherol [74, 75].

Interestingly, in *Ginkgo biloba* seedlings, treatment with 100 mmol/L NaCl was found to improve the biosynthesis of flavonoids and flavonol [76]. In addition, the upregulated expression of five flavonoid biosynthesis-related genes was also detected. The authors indicated that *G. biloba* seedlings could tolerate low-level soil salinity stress via the regulation of flavonoid biosynthesis, which was accelerated in response to environmental stress.

2.2.4 Tocopherols

Tocopherols constitute a group of lipophilic compounds that are important antioxidant and nonenzymatic components capable of inhibiting lipid peroxidation. Tocopherols are also essential for maintaining membrane integrity. These lipophilic compounds are synthesized from homogentisic acid and isopentenyl diphosphate in the plastid envelope. Tocopherol exists in four isomeric forms (α-, β-, γ-, and δ-), with the predominant forms being α- and γ-tocopherol. The composition of tocopherol depends on the genotypic features of the plant, growth conditions, tissue type, and stress intensity. Tocopherols are present in seeds, fruits, roots, tubers, cotyledons, hypocotyls, stems, leaves, and flowers. In addition, α-tocopherol may regulate the concentrations of plant hormones, such as jasmonic acid, which control both growth and development [38, 77].

As a component of thylakoid membranes, tocopherol acts as an important scavenger of 1O2 and OH while maintaining a stable redox status and the structure and function of PSII. Tocopherols have also been found to reduce lipid peroxyl radicals (obtained from lipid peroxidation) to their corresponding hydroperoxides [78]. In addition, tocopherols participate in cell signaling and they may protect embryos during germination from ROS, senescence, and stress [79, 80]. Recent studies have also shown that tocopherol plays a positive role under salt-stress conditions by controlling Na+/K+ homeostasis and the hormonal balance while minimizing oxidative stress [81]. Additionally, it has been demonstrated that the foliar application of a-tocopherol in onion plants and *Vicia fava* improve salt tolerance with an increase in plant growth and productivity in this condition [82, 83].

2.2.5 Carotenoids

Carotenoids are hydrophobic pigments derived from isoprene that play essential roles in photosynthesis and nutrition and protect against photo-oxidative damage in higher plants [37]. Carotenoids are mainly present in the form of 40-carbon tetraterpene, which consists of eight isoprene units [84]. In plants, carotenoids are synthesized in plastids from isoprenoid precursors. Lycopene is a principal carotenoid and is a product of the sequential desaturations of phytoene and may be converted into b-carotene by lycopene b-cyclase (Lcy-b) [85, 86]. Carotenoids act as light harvesters by dissipating excess energy as heat, and they also protect the photosynthetic apparatus from the free radicals produced during photosynthesis by stabilizing photosystem I and II and the thylakoid membrane. Furthermore, carotenoids are precursors of hormones like abscisic acid and strigolactones [87].

Carotenoids quench the O2 and H2O2 formed under salt stress due to their low triplet state energy and in this way, protect lipids from peroxidation while suppressing radical chain reactions [88–90]. Li et al. [91] recently showed that the overexpression of three genes associated with carotenoid biosynthesis from *Lycium chinenses* enhanced salt tolerance in tobacco plants and suggested that plants may synthesize and accumulate more carotenoids under salt-stress conditions, showing higher resistance to oxidative stress, by an increase of the expression level and activity of antioxidant enzymes.

3. Osmolytes that maintain cell turgor

Salinity decreases the osmotic potential of soil, which in turn limits water absorption by plants. In addition, the toxicity of Na + and Cl- impedes the uptake of essential nutrients. In particular, Na + toxicity can disrupt the absorption of K+ and Ca2+, favoring the production of ROS and inducing oxidative damage [92].

Plants have developed responses to deal with water loss and the transport and accumulation of toxic ions. Given that water loss in plants originates due to the low osmotic potential of the soil, plants must employ compensatory mechanisms [93]. Plant cells permit higher influxes of Na+, which decrease the water potential and cause osmotic stress in cells. Salts are translocated from the roots to shoots to maintain ion homeostasis, although this reduces photosynthetic efficiency due to the inhibition of photosystem II (PSII) [94]. To cope with osmotic stress and maintain ion homeostasis, plants biosynthesize osmolytes.

Osmolytes or compatible solutes are small molecules with low molecular weights that are electrically neutral, highly soluble, and do not affect normal biochemical processes [95]. The biosynthesis and accumulation of osmolytes constitutes an essential protection strategy for plants under abiotic stress. Multiple studies have elucidated the roles that osmolytes play in abiotic stress tolerance due to natural accumulation [96, 97], exogenous application [98, 99], or the transgenic expression of osmolyte pathway genes [100, 101], which are summarized in **Table 1**.

	Type of study	Species	Reference
Amino acids			
Proline	Exogenous application	*Solanum lycopersicum*, *Glycine max*, *Pisum sativum*	[102–104]
GABA	Natural accumulation	*Nicotiana tabacum and Sorghum bicolor*	[105, 106]
Polyamines			
Putrescine	Natural accumulation	*Cajanus cajan*	[107]
Spermidine	Exogenous application	*Oryza sativa*	[108]
Spermine	Natural accumulation	*O. sativa*	[109]
Betaines			
Glycine betaine	Exogenous application	*O. sativa*, *Glicine max*, and *Phaseolus vulgaris*	[110–112]
Sugars			
Glucose and fructose	Natural accumulation	*Vitis vinifera* and *Lepidium crassifolium*	[113, 114]
Trehalose	Transgenic expression	*O. sativa*	[115]
Raffinose	Transgenic expression	*Arabidopsis thaliana*	[116]
Polyols			
Mannitol	Transgenic expression	*N. tabacum*	[117]
Myo-inositol and pinitol	Transgenic expression	*N. tabacum*	[118, 119]
GABA = Gamma amino butyric acid			

Table 1.
Studies in plants that directly correlate metabolite accumulation to increased salinity stress tolerance.

The primary function of osmolytes is to accumulate inside the cell to maintain the osmotic balance between the plant cell and its surroundings [120]. Osmolytes can also act as chemical chaperones by stabilizing and protecting proteins and membranes [121] or by acting as ROS scavengers [122]. Some of these compatible solutes are highly effective in reducing the extent of K+ loss in response to salt stress [123]. Overall, osmolytes are chemically diverse, although they can be broadly grouped into amino acids, polyamines, betaines, sugars, and polyols, which are shown in **Figure 2**.

3.1 Amino acids

Amino acids are the building blocks of proteins, which carry out many biological processes and provide structure and support for cells [124]. In addition, many studies have reported the accumulation of amino acids in plants exposed to drought and salinity conditions [97, 98, 125, 126]. This accumulation may be due to elevated amino acid production and/or the stress-induced breakdown of proteins [127]. Some of the amino acids that have been found to accumulate are alanine, arginine, glycine, serine, leucine, valine, and proline, in addition to non-protein amino acids like citrulline, ornithine, and gamma amino butyric acid (GABA) [128, 129].

Proline is the most important osmolyte and signaling molecule and generally accumulates in the cytosol. Proline also acts as an ROS scavenger and molecular chaperone, stabilizing the structure of proteins and protecting cells from potential damage induced by osmotic and oxidative stress [120, 130, 131]. Proline is mainly synthesized from glutamate by Δ1-pyrroline-5-carboxylate (P5C) synthetase and P5C reductase enzymes, although under nitrogen-limited conditions, proline is

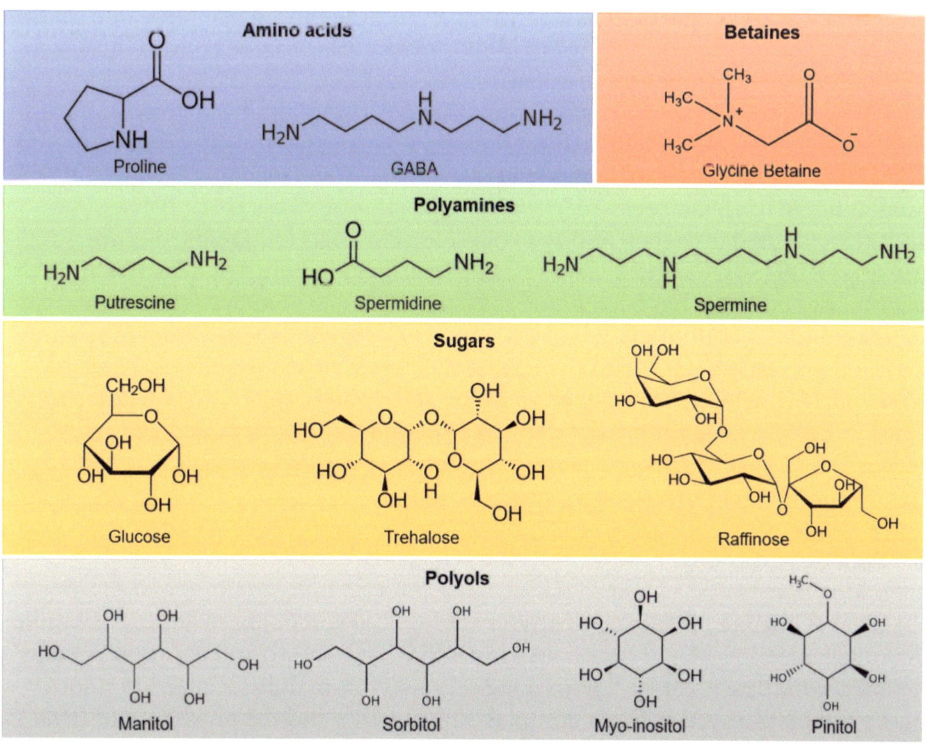

Figure 2.
Chemical structures of the most common osmolytes observed in salt-stress responses in plants. These structures can be grouped into amino acids, polyamines, betaines, sugars, and polyols.

synthesized in the ornithine pathway and transformed into P5C/GSA (glutamate-1-semialdehyde) via ornithine-δ-aminotransferase [132, 133].

Proline has been associated with increased salt tolerance in tomato (*Solanum lycopersicum*) [102], soybean (*Glycine max*) [103], groundnut (*Arachis hypogaea*) [134], pea (*Pisum sativum*) [104], sainfoin (*Onobrychis viciaefolia*), and mung bean (*Vigna radiata*) plants [100]. Researchers have also reported that the exogenous application of proline enhances its endogenous level, thus promoting growth and the antioxidant defense system and decreasing the uptake rate of Na + and Cl- [99, 135].

Gamma amino butyric acid is a four-carbon non-protein amino acid with an amino group attached to γ-carbon instead of α- carbon. GABA is mainly synthesized from glutamate in the cytosol by glutamate decarboxylase and then transported to the mitochondria [136]. GABA metabolism has been associated with the carbon/nitrogen balance and ROS scavenging [137, 138]. In addition, GABA accumulation and metabolism have been found to be activated by salt exposure in tobacco plants [105]. Moreover, GABA-T *Arabidopsis* mutants have been found to be hypersensitive to ionic stress, showing elevated levels of amino acids (including GABA) [139]. In sorghum, GABA may also contribute to CSF20 osmoregulation and signaling to increase salt tolerance [106].

3.2 Polyamines

Polyamines (PAs) are low molecular weight aliphatic nitrogenous bases containing two or more amino groups [140]. The most common PAs in higher plants are putrescine, spermidine, and spermine. Putrescine can be produced from ornithine by ornithine decarboxylase or from arginine by arginine decarboxylase. Putrescine is then converted to spermidine (spermidine synthase) and spermidine into spermine (spermine synthase). Both spermidine and spermine can be converted back to putrescine by polyamine-oxidases [36].

PAs regulate diverse cellular functions that are essential for cell growth, including senescence, development, cell proliferation, and signal transduction while also regulating the expression of genes in response to various stressors. However, PA accumulation has been detected in plants under abiotic stress [141]. Polyamines act as osmolytes due to their ability to block ion channels (cationic structures) and scavenge ROS [36]. Exogenous spermidine treatment in both salt-sensitive and salt-tolerant rice cultivars has been found to result in plasma membrane recovery after injury induced by salinity [108]. In addition, the exogenous application of spermidine and putrescine was found to increase the postharvest shelf-life of *Capsicum annuum* [142], while improving grain filling and drought tolerance in wheat plants [143]. It has also been reported that putrescine exhibits salinity tolerance in the pigeon pea (*Cajanus cajan*) by modulating the anabolic and catabolic enzyme activities responsible for putrescine biosynthesis [107].

3.3 Betaines

Glycine betaine (GB) is a quaternary ammonium compound whose distribution among plants is restricted to certain species like *Arabidopsis*, and many crop plants do not accumulate it [144]. This organic compound is mainly localized in chloroplasts and plays a vital role in chloroplast adjustments and the protection of thylakoid membranes, which helps to maintain the photosynthetic efficiency of the plant [145, 146]. At the same time, GB encourages water to flow into cells, which helps to maintain intracellular osmotic equilibrium and regulates the signaling transduction cascade under stressful conditions [147]. GB is synthesized in chloroplasts from serine via ethanolamine, choline, and betaine aldehyde. Choline is converted to betaine aldehyde by choline monooxygenase, which is then converted to GB by betaine aldehyde dehydrogenase [148].

The exogenous application of GB mitigates the adverse effects of salinity stress in some plant species. For example, the foliar application of GB was found to result in significantly improved salt tolerance in rice plants [110], and the exogenous application of GB on tomato plants subjected to salt stress resulted in an increase in fruit yield of ~40% compared with that of untreated plants [149]. Salinity tolerance in response to GB has also been observed in mung bean (*V. radiata*) [150], green bean (*Phaseolus vulgaris*) [108], and soybean (*G. max*) plants [111].

3.4 Sugars

Sugars provide carbon and energy for cellular metabolic processes while regulating plant growth and development. However, under stressful conditions, carbohydrate metabolism results in an increase in sugar levels. The production and collection of soluble sugars directly contributes to ROS scavenging, osmotic adjustments, carbon storage, and the stabilization of protein structures like Ribulose-1,5-bisphosphate carboxylase–oxygenase [151]. Sugars are also the main osmolytes that participate in osmotic adjustments and can contribute up to 50% of the total osmotic potential in some plant species [152]. Structurally, all kinds of sugars have been reported in response to salinity, including monosaccharides (glucose and fructose), disaccharides (trehalose and sucrose), and oligosaccharides (raffinose family) (**Table 1**).

Studies have shown that plants mobilize starch and fructans from storage organs (roots, stems, and amyloplasts in leaves) to increase the accumulation of sugars, such as glucose, fructose, and sucrose. Starch hydrolysis by the b-amolytic pathway represents the primary means of starch degradation in leaves under normal growth conditions and may also be involved in stress-induced starch hydrolysis. Downton [113] shows that *Vitis vinifera* salt-stressed leaves contained decreased amounts of sucrose and starch but increased levels of reducing sugars. Rathert [153] showed that salt stress induced the restricted utilization of leaf sucrose but not of foliar starch in *G. max* varieties. In addition, Murakeözy et al. [114] found a high accumulation of soluble carbohydrates in *Lepidium crassifolium* (Brassicaceae) leaves in response to salinity.

Similarly, trehalose accumulates in many organisms due to various abiotic stressors and has been reported to act as an osmolyte [154]. Trehalose is a disaccharide that is synthesized by a two-step process in which trehalose-6-phosphate is first synthesized from glucose-6-phosphate and then dephosphorylated to trehalose by trehalose-6-phosphate phosphatase [155]. Several studies have evaluated the transgenic expression of trehalose biosynthesis genes, which have been shown to enhance trehalose metabolism and tolerance to abiotic stress. For instance, the overexpression of otsA and otsB in transgenic rice plants was found to result in the increased accumulation of trehalose content and an overall improved photosynthetic capacity, reducing oxidative damage and improving ion uptake and partitioning under conditions of salt stress [115]. Also, the induction of OsTPP1 and OsTPP2 by the exogenous application of ABA was found to enhance cold, salinity, and drought tolerance in rice plants [156].

The raffinose family oligosaccharides (RFO), such as raffinose, stachyose, and verbascose, are macromolecules that act to protect plants from drought, salt, cold, freezing, and oxidative stress [157]. In addition, RFOs have been implicated in membrane protection and radical scavenging. Furthermore, Arabidopsis plants overexpressing Arabidopsis GolS1 or GolS2 were found to accumulate high levels of galactinol and raffinose and were more tolerant to salinity stress [116].

3.5 Polyols

Polyols or sugar alcohols are polyhydric alcohols that are widely distributed in the plant kingdom, and they can be grouped into linear and cyclic structures. The most common polyols are mannitol, glycerol, sorbitol (linear), and myo-inositol, along with the methylated derivatives ononitol and pinitol (cyclic) [158, 159]. Polyols accumulate in the cytoplasm and act as osmoprotectants and ROS scavengers, preventing water loss and oxidative damage to membranes and enzymes. Polyols originate via the reduction of aldoses or their phosphate esters and are generally water soluble in nature [160].

Williamson [161] reported that mannitol, sorbitol, and inositol increase drought and salinity tolerance in some plants. Transgenic tobacco plants overexpressing bacterial mannitol-1-phosphate dehydrogenase were found to accumulate high concentrations of mannitol in their leaves and roots and exhibited a high degree of salt tolerance [101]. Moreover, Adams et al. [162] reported that myo-inositol, ononitol, and pinitol accumulated under salt-stress conditions in several halotolerant plant species. Furthermore, the over-expression of MIPS and IMT from halotolerant plants was found to increase cyclic polyol levels and salt-stress tolerance in tobacco plants [118, 119].

4. Conclusions and perspectives

Salt stress is one of the most important factors that limits the productivity of agricultural crops and threatens global food security. Salinity produces osmotic stress and ionic toxicity in plants, which alters cell homeostasis and reduces plant water absorption, adversely affecting growth and plant productivity. Osmotic imbalances have been found to cause an overproduction of ROS that leads to the oxidation of biomolecules, cell membrane damage, protein degradation, enzyme inhibition, DNA mutations, and cellular death. To limit the damage, cells turn on enzymatic and nonenzymatic antioxidant machinery. The ROS detoxification system is very complex and controlled at multiple levels because ROS participate in biological processes like photosynthesis, signal transduction, development, and programmed cell death. Therefore, to evaluate the negative and positive roles of ROS, it is important to understand tolerance mechanisms in plants and the ability to control or moderate ROS levels. Recently, it has been proposed that ROS homeostasis may be modulated through the use of nanoparticles that activate the expression of antioxidant enzyme genes or that have the ability to scavenge ROS [22], thus improving salt tolerance.

Another response to salinity in plants is the induction of osmolyte biosynthesis to stave off water loss and the accumulation of toxic ions. The production and accumulation of osmolytes inside the cell is essential for the protection and survival of plants under salt stress and other forms of environmental stress, although their production is taxonomically restricted. It has been shown that the exogenous application of osmolytes, such as aminoacids, betaines, or sugars, and the overexpression of osmolyte genes constitute viable options to improve plant resistance to salinity. Thus, the use and application of osmolytes in agricultural settings should continue to expand. Current studies have shown that when taken together, metabolic and physiological plant responses can provide valuable information of the possible mechanisms required for plants to adapt to stress. Thus, modulating one component of the antioxidative defense system or a particular metabolite might be insufficient to confer resistance to the entire plant. To generate salt-tolerant species, it is essential to use modified plants that incorporate more than one component, such as the use of multiple antioxidants or metabolites or a combination of both. In this way, further progress in genomics, proteomics, and metabolomics, together with improvements in biotechnological tools, will allow for the underlying mechanisms of salt-tolerant species to be fully elucidated and understood.

Acknowledgements

We would like to thank Programa para el Desarrollo Profesional Docente (PRODEP), University of Guadalajara, Consejo Nacional de Ciencia y Tecnología (CONACYT), and Laboratorio Nacional PlanTECC for the use of facilities and the support provided via the project 315918.

Author details

Julio Armando Massange-Sánchez[1], Carla Vanessa Sánchez-Hernández[2], Rosalba Mireya Hernández-Herrera[2] and Paola Andrea Palmeros-Suárez[2]*

1 Centro de Investigación y Asistencia en Tecnología y Diseño del Estado de Jalisco A.C. Guadalajara, Jalisco, México

2 Centro Universitario de Ciencias Biológicas y Agropecuarias, University of Guadalajara, Zapopan, Jalisco, México

*Address all correspondence to: paola.palmeros@academicos.udg.mx

References

[1] Teakle NL, Tyerman SD. Mechanisms of Cl- transport contributing to salt tolerance. Plant Cell Environ. 2010;33(4):566-89. DOI:10.1111/j.1365-3040.2009.02060.x

[2] García-Caparrós P, Hasanuzzaman M, Lao MT. Ion homeostasis and antioxidant defense toward salt tolerance in plants. In: Hasanuzzaman M, Fujita M, Oku H, Nahar K, Hawrylak-Nowak B, editors. Plant Nutrients and Abiotic Stress Tolerance. 1 st ed. Singapore: Springer; 2018. p. 415-36. DOI:10.1007/978-981-10-9044-8_18

[3] Munns R. Approaches to identifying genes for salinity tolerance and the importance of timescale. In: Sunkar R, editor. Plant Stress Tolerance: Methods and Protocols. 1 st ed. Totowa, NJ: Humana Press; 2010. p. 25-38. DOI:10.1007/978-1-60761-702-0_2

[4] Kaashyap M, Ford R, Kudapa H, Jain M, Edwards D, Varshney R, et al. Differential regulation of genes involved in root morphogenesis and cell wall modification is associated with salinity tolerance in chickpea. Sci Rep. 2018;8(1):4855. DOI: 10.1038/s41598-018-23116-9

[5] de Azevedo Neto AD, Prisco JT, Enéas-Filho J, Abreu CEB de, Gomes-Filho E. Effect of salt stress on antioxidative enzymes and lipid peroxidation in leaves and roots of salt-tolerant and salt-sensitive maize genotypes. Environ Exp Bot. 2006;56(1):87-94.

[6] Athar HR, Ashraf M. Strategies for crop improvement against salinity and drought stress: An Overview. Salinity and Water Stress. In: Ashraf M, Ozturk M, Athar HR, editors. Improving Crop Efficiency. 1 st ed. Dordrecht: Springer; 2009. p. 1-16. DOI:10.1007/978-1-4020-9065-3_1

[7] Shrivastava P, Kumar R. Soil salinity: A serious environmental issue and plant growth promoting bacteria as one of the tools for its alleviation. Saudi J Biol Sci. 2015;22(2):123-31.

[8] Munns R, Tester M. Mechanisms of salinity tolerance. Annu Rev Plant Biol. 2008;59(1):651-81. DOI:10.1146/annurev.arplant.59.032607.092911

[9] Khataar M, Mohammadi MH, Shabani F. Soil salinity and matric potential interaction on water use, water use efficiency and yield response factor of bean and wheat. Sci Rep. 2018;8(1): 2679. DOI:10.1038/s41598-018-20968-z

[10] Shahbaz M, Mushtaq Z, Andaz F, Masood A. Does proline application ameliorate adverse effects of salt stress on growth, ions and photosynthetic ability of eggplant (*Solanum melongena* L.). Sci Hortic. 2013;164:507-11.

[11] Munns R. Comparative physiology of salt and water stress. Plant Cell Environ. 2002;25(2):239-50. DOI:10.1046/j.0016-8025.2001.00808.x

[12] Rahnama A, Poustini K, Tavakkol-Afshari R, Ahmadi A, Alizadeh H. Growth properties and ion distribution in different tissues of bread wheat genotypes (*Triticum aestivum* L.) differing in salt tolerance. J Agron Crop Sci. 2011;197(1):21-30. DOI:10.1111/j.1439-037X.2010.00437.x

[13] Parida AK, Das AB. Salt tolerance and salinity effects on plants: a Review. Ecotoxicol Environ Saf. 2005;60(3): 324-49.

[14] Acosta-Motos JR, Ortuño MF, Bernal-Vicente A, Diaz-Vivancos P, Sanchez-Blanco MJ, Hernandez JA. Plant responses to salt stress: Adaptive mechanisms. Agronomy. 2017; 7(1):18 DOI: 10.3390/agronomy7010018

[15] Rasool S, Hameed A, Azooz MM, Muneeb-u-Rehman, Siddiqi TO, Ahmad P. Salt Stress: Causes, Types and Responses of Plants. In: Ahmad P, Azooz MM, Prasad MN V, editors. Ecophysiology and Responses of Plants under Salt Stress. 1 st ed. New York: Springer; 2013; p. 1-24.DOI: 10.1007/978-1-4614-4747-4_1

[16] Pandey P, Irulappan V, Bagavathiannan M V, Senthil-Kumar M. Impact of combined abiotic and biotic stresses on plant growth and avenues for crop improvement by exploiting physio-morphological traits. Front Plant Sci. 2017; 18(8): 537. DOI: 10.3389/fpls.2017.00537.

[17] Eynard A, Lal R, Wiebe K. Crop response in salt-affected soils. J Sustain Agric. 2005;27(1):5-50. DOI:10.1300/J064v27n01_03

[18] García-Caparrós P, Hasanuzzaman M, Lao MT. Oxidative stress and antioxidant defense in plants under salinity. In: Hasanuzzaman M, Fotopoulos V, Nahar K, Fujita M editors. Reactive oxygen, nitrogen and sulfur species in plants. 1 st ed. Chichester: Wiley 2019; p. 291-309. DOI:10.1002/9781119468677.ch12

[19] Yang Y, Guo Y. Unraveling salt stress signaling in plants. J Integr Plant Biol. 2018;60(9):796-804. DOI:10.1111/jipb.12689

[20] Ahanger MA, Tomar NS, Tittal M, Argal S, Agarwal RM. Plant growth under water/salt stress: ROS production; antioxidants and significance of added potassium under such conditions. Physiol Mol Biol Plants.2017;23(4): 731-44.

[21] Ismail A, El-Sharkawy I, Sherif S. Salt stress signals on demand: Cellular events in the right context. Int J Mol Sci. 2020;21(11):1-22.

[22] Liu J, Fu C, Li G, Nauman Khan M, Wu H. Ros homeostasis and plant salt tolerance: plant nanobiotechnology updates. Sustain. 2021;13(6).

[23] Mittler R, Vanderauwera S, Gollery M, Van Breusegem F. Reactive oxygen gene network of plants. Trends Plant Sci. 2004;9(10):490-8.

[24] Mittler R, Vanderauwera S, Suzuki N, Miller G, Tognetti VB, Vandepoele K, et al. ROS signaling: the new wave?. Trends Plant Sci. 2011; 16(6):300-9.

[25] Jacoby RP, Taylor NL, Millar AH. The role of mitochondrial respiration in salinity tolerance. Trends Plant Sci. 2011;16(11):614-23.

[26] Torres MA, Jones JDG, Dangl JL. Reactive oxygen species signaling in response to pathogens. Plant Physiol. 2006;141(2):373-8. DOI: 10.1104/pp.106.079467

[27] Abogadallah GM. Insights into the significance of antioxidative defense under salt stress. Plant Signal Behav. 2010;5(4):369-74. DOI: 10.4161/psb.5.4.10873

[28] Hirayama T, Shinozaki K. Research on plant abiotic stress responses in the post-genome era: past, present and future. Plant J. 2010;61(6):1041-52. DOI:10.1111/j.1365-313X.2010.04124.x

[29] Hasanuzzaman M, Bhuyan MHMB, Anee TI, Parvin K, Nahar K, Mahmud J Al, et al. Regulation of ascorbate-glutathione pathway in mitigating oxidative damage in plants under abiotic stress. Antioxidants. 2019;8(9):384.

[30] Hossain MS, Dietz K-J. Tuning of Redox Regulatory Mechanisms, Reactive Oxygen Species and Redox Homeostasis under Salinity Stress. Front Plant Sci. 2016; 7: 548.

[31] Teotia S, Singh D. Oxidative stress in plants and its management. In: Gaur RK, Sharma P, editors. Approaches to plant

stress and their management. 1 st. ed. New Delhi: Springer; 2014. p. 227-53. DOI:10.1007/978-81-322-1620-9_13

[32] Kibria MG, Hoque MA. A Review on plant responses to soil salinity and amelioration strategies. Open J Soil Sci. 2019;09(11):219-31.

[33] Ahmad R, Hussain S, Anjum MA, Khalid MF, Saqib M, Zakir I, et al. Oxidative stress and antioxidant defense mechanisms in plants under salt stress. In: Hasanuzzaman M, Hakeem KR, Nahar K, Alharby HF, editors. Plant abiotic stress tolerance: Agronomic, molecular and biotechnological Approaches. 1 st ed. Cham: Springer; 2019. p. 191-205. DOI:10.1007/978-3-030-06118-0_8

[34] Bose J, Rodrigo-Moreno A, Shabala S. ROS homeostasis in halophytes in the context of salinity stress tolerance. J Exp Bot. 2014;65(5): 1241-57.

[35] Dvořák P, Krasylenko Y, Zeiner A, Šamaj J, Takáč T. Signaling toward reactive oxygen species-scavenging enzymes in plants. Front Plant Sci. 2021; 11: 2178.

[36] Gill SS, Tuteja N. Polyamines and abiotic stress tolerance in plants. Plant Signal Behav. 2010;5(1):26-33.

[37] Uarrota VG, Stefen DLV, Leolato LS, Gindri DM, Nerling D. Revisiting carotenoids and their role in plant stress responses. In: Gupta DK, Palma JM, Corpas FJ, editors. Biosynthesis to Plant Signaling Mechanisms During Stress. Antioxidants and antioxidant enzymes in higher plants. 1 st ed. Cham: Springer; 2018. p. 207-32. DOI:10.1007/978-3-319-75088-0_10

[38] Štolfa Čamagajevac I, Žuna Pfeiffer T, Špoljarić Maronić D. Abiotic stress response in plants: The relevance of tocopherols. In: Gupta DK, Palma JM, Corpas FJ, editors. Antioxidants and antioxidant enzymes in higher plants. 1 st ed. Cham: Springer; 2018. p. 233-51. DOI:10.1007/978-3-319-75088-0_11

[39] Bowler C, Van Camp W, Van Montagu M, Inzé D, Asada K. Superoxide Dismutase in Plants. CRC Crit Rev Plant Sci. 1994;13(3):199-218. DOI:10.1080/07352689409701914

[40] Yan H, Li Q, Park S-C, Wang X, Liu Y, Zhang Y, et al. Overexpression of CuZnSOD and APX enhance salt stress tolerance in sweet potato. Plant Physiol Biochem. 2016;109:20-7.

[41] Mhamdi A, Noctor G, Baker A. Plant catalases: Peroxisomal redox guardia. Arch. Biochem. Biophys. 2012; 525(2):181-194. DOI: 10.1016/j.abb.2012.04.015

[42] Corpas FJ, Chaki M, Fernández-Ocaña A, Valderrama R, Palma JM, Carreras A, et al. Metabolism of reactive nitrogen species in pea plants under abiotic stress conditions. Plant Cell Physiol. 2008;49(11):1711-22. DOI: 10.1093/pcp/pcn144

[43] Hussain S, Rao MJ, Anjum MA, Ejaz S, Zakir I, Ali MA, et al. Oxidative stress and antioxidant defense in plants under drought conditions. In: Hasanuzzaman M, Hakeem KR, Nahar K, Alharby HF, editors. Plant Abiotic Stress Tolerance: Agronomic, Molecular and Biotechnological Approaches. 1 st ed. Cham: Springer; 2019. p. 207-19. DOI:10.1007/978-3-030-06118-0_9

[44] Gondim FA, Gomes-Filho E, Costa JH, Mendes Alencar NL, Prisco JT. Catalase plays a key role in salt stress acclimation induced by hydrogen peroxide pretreatment in maize. Plant Physiol Biochem. 2012;56:62-71.

[45] Anjum NA, Sharma P, Gill SS, Hasanuzzaman M, Khan EA, Kachhap K, et al. Catalase and ascorbate peroxidase—representative

H2O2-detoxifying heme enzymes in plants. Environ Sci Pollut Res. 2016; 23(19):19002-29. DOI:10.1007/s11356-016-7309-6

[46] Rohman MM, Islam MR, Naznin T, Omy SH, Begum S, Alam SS, et al. Maize Production under salinity and drought conditions: Oxidative stress regulation by antioxidant defense and glyoxalase systems. In: Hasanuzzaman M, Hakeem KR, Nahar K, Alharby HF, editors. Plant Abiotic Stress Tolerance: Agronomic, Molecular and Biotechnological Approaches. 1 st ed. Cham: Springer; 2019. p. 1-34. DOI: 10.1007/978-3-030-06118-0_1

[47] Chew O, Whelan J, Millar AH. Molecular definition of the ascorbate-glutathione cycle in Arabidopsis mitochondria reveals dual targeting of antioxidant defenses in Plants. J Biol Chem. 2003;278(47):46869-77.

[48] Latowski D, Surówka E, Strzałka K. Regulatory role of components of ascorbate–glutathione pathway in Plant Stress Tolerance. In: Anjum NA, Chan M-T, Umar S, editors. Ascorbate-Glutathione Pathway and Stress Tolerance in Plants. 1 st ed. Dordrecht: Springer; 2010. p. 1-53. DOI:10.1007/978-90-481-9404-9_1

[49] Martínez JP, Araya H. Ascorbate–glutathione cycle: Enzymatic and non-enzymatic integrated mechanisms and its biomolecular regulation. In: Anjum NA, Chan M-T, Umar S, editors. Ascorbate-Glutathione Pathway and Stress Tolerance in Plants. 1 st ed. Dordrecht: Springer; 2010. p. 303-22. DOI:10.1007/978-90-481-9404-9_11

[50] Gaafar RM, Seyam MM. Ascorbate-glutathione cycle confers salt tolerance in Egyptian lentil cultivars. Physiol Mol Biol Plants. 2018;24(6):1083-92.

[51] Li H, Wang H, Wen W, Yang G. The antioxidant system in *Suaeda salsa* under salt stress. Plant Signal Behav. 2020;15(7):1771939. DOI:10.1080/15592324.2020.1771939

[52] Kim YS, Kim IS, Shin SY, Park TH, Park HM, Kim YH, et al. Overexpression of dehydroascorbate reductase confers enhanced tolerance to salt stress in rice plants (*Oryza sativa L. japonica*). J Agron Crop Sci. 2014;200(6):444-56. DOI:10.1111/jac.12078

[53] Gest N, Gautier H, Stevens R. Ascorbate as seen through plant evolution: the rise of a successful molecule? J Exp Bot. 2013;64(1):33-53. DOI:10.1093/jxb/ers297

[54] Smirnoff N. Ascorbate biosynthesis and function in photoprotection. Phil. Trans. R. Soc. Lond B. 2000;355(1402): 1455-64.

[55] Yoshimura K, Ishikawa T. Chemistry and metabolism of ascorbic acid in plants In: Hossain MA, Munné-Bosch S, Burritt DJ, Diaz-Vivancos P, Fujita M, Lorence A, editors. Ascorbic acid in plant growth, Development and Stress Tolerance. 1 st ed. Cham: Springer; 2017. p. 1-23.DOI: 10.1007/978-3-319-74057-7_1

[56] Gallie DR. The role of l-ascorbic acid recycling in responding to environmental stress and in promoting plant growth. J Exp Bot. 2013;64(2): 433-43. DOI:10.1093/jxb/ers330

[57] Hernández JA, Barba-Espín G, Clemente-Moreno MJ, Díaz-Vivancos P. Plant responses to salinity through an antioxidative metabolism and proteomic point of view In: Sarwat M, Ahmad A, Abdin MZ, Ibrahim MM, editors. Stress Signaling in Plants: Genomics and Proteomics Perspective, Volume 2. 1 st ed. Cham: Springer; 2017. p. 173-200. DOI.10.1007/978-3-319-42183-4_8

[58] Bulley S, Laing W. The regulation of ascorbate biosynthesis. Curr Opin Plant Biol. 2016;33:15-22.

[59] Aliniaeifard S, Hajilou J, Tabatabaei SJ, Sifi-Kalhor M. Effects of ascorbic acid and reduced glutathione on the alleviation of salinity stress in olive plants. Int J Fruit Sci. 2016; 16(4):395-409.

[60] Sadak M. Impacto de las aplicaciones foliares de ácido ascórbico y α- tocoferol en la actividad antioxidante y algunos aspectos bioquímicos de cultivares de lino sometidos a estrés por salinidad. Acta Biol Colom. 2015; 20(2):209-222.

[61] Noctor G, Mhamdi A, Chaouch S, Han YI, Neukermans J, Márquez-García B, et al. Glutathione in plants: an integrated overview. Plant Cell Environ. 2012;35(2):454-84. DOI:10.1111/j.1365-3040.2011.02400.x

[62] Mahmood Q, Ahmad R, Kwak S-S, Rashid A, Anjum NA. Ascorbate and Glutathione: Protectors of plants in oxidative stress. In: Anjum NA, Chan M-T, Umar S, editors. Ascorbate-Glutathione Pathway and Stress Tolerance in Plants. 1 st ed. Dordrecht: Springer; 2010. p. 209-29. DOI: 10.1007/978-90-481-9404-9_7

[63] Mullineaux PM, Rausch T. Glutathione, photosynthesis and the redox regulation of stress-responsive gene expression. Photosynth Res. 2005;86(3):459-74. DOI:10.1007/s11120-005-8811-8

[64] Sharma P, Jha AB, Dubey RS, Pessarakli M. Reactive oxygen species, oxidative damage, and antioxidative defense mechanism in plants under stressful conditions. Polle A, editor. J Bot. 2012;2012:217037. DOI: 10.1155/2012/217037

[65] Foyer CH, Noctor G. Oxidant and antioxidant signalling in plants: a re-evaluation of the concept of oxidative stress in a physiological context. Plant Cell Environ. 2005;28(8):1056-71. DOI:10.1111/j.1365-3040.2005.01327.x

[66] Labudda M, Azam FMS. Glutathione-dependent responses of plants to drought: A review. Acta Soc Bot Pol. 2014;83(1):3-12.

[67] Liebthal M, Dietz K-J. The fundamental role of reactive oxygen species in plant stress response. In: Sunkar R, editor. Plant Stress Tolerance: Methods and Protocols. 1 st ed. New York: Springer; 2017. p. 23-39. DOI:10.1007/978-1-4939-7136-7_2

[68] Hussain BM, Akram S, Raffi SA, Burritt DJ, Hossain MA. Exogenous glutathione improves salinity stress tolerance in rice (*Oryza sativa* L.). Plant Gene Trait. 2016; 7(11):1-17 DOI: 10.5376/pgt.2016.07.0011

[69] Zhou Y, Wen Z, Zhang J, Chen X, Cui J, Xu W, Li H. Exogenous glutathione alleviates salt-induced oxidative stress in tomato seedlings by regulating glutathione metabolism, redox status, and the antioxidant system. Sci Hortic. 2017;2020:90-101

[70] Djoukeng JD, Arbona V, Argamasilla R, Gomez-Cadenas A. Flavonoid profiling in leaves of citrus genotypes under different environmental situations. J Agric Food Chem. 2008;56(23):11087-97. DOI:10.1021/jf802382y

[71] Di Ferdinando M, Brunetti C, Fini A, Tattini M. Flavonoids as antioxidants in plants under abiotic stresses. In: Ahmad P, Prasad MN V, editors. Abiotic Stress Responses in Plants: Metabolism, Productivity and Sustainability. 1 st ed. New York: Springer; 2012. p. 159-79. DOI:10.1007/978-1-4614-0634-1_9

[72] Agati G, Azzarello E, Pollastri S, Tattini M. Flavonoids as antioxidants in plants: Location and functional significance. Plant Sci. 2012;196:67-76.

[73] Kumar S, Pandey AK. Chemistry and biological activities of flavonoids:

An overview. Sci World J. 2013;2013:162750. DOI:10.1155/2013/162750

[74] Mierziak J, Kostyn K, Kulma A. Flavonoids as important molecules of plant interactions with the environment. Molecules. 2014;19(10):16240-65.

[75] Hernández I, Alegre L, Van Breusegem F, Munné-Bosch S. How relevant are flavonoids as antioxidants in plants? Trends Plant Sci. 2009;14(3): 125-32.

[76] Xu N, Liu S, Lu Z, Pang S, Wang L, Wang L, Li W. Gene expression profiles and flavonoid accumulation during salt stress in *Ginkgo biloba* seedlings. Plants 2020; 9(9):1162. DOI:10.3390/plants9091162

[77] Munné-Bosch S, Alegre L. The function of tocopherols and tocotrienols in plants. CRC Crit Rev Plant Sci. 2002;21(1):31-57. DOI:10.1080/0735-260291044179

[78] Maeda H, Song W, Sage TL, DellaPenna D. Tocopherols play a crucial role in low-temperature adaptation and phloem loading in Arabidopsis. Plant Cell. 2006;18(10):2710-32.

[79] Krieger-Liszkay A, Trebst A. Tocopherol is the scavenger of singlet oxygen produced by the triplet states of chlorophyll in the PSII reaction centre. J Exp Bot. 2006;57(8):1677-84. DOI:0.1093/jxb/erl002

[80] Bafeel SO, Ibrahim MM. Antioxidants and accumulation of α-tocopherol induce chilling tolerance in *Medicago sativa*. Int J Agric Biol. 2008;10(6):593-8.

[81] Ellouzi H, Hamed K Ben, Cela J, Müller M, Abdelly C, Munné-Bosch S. Increased sensitivity to salt stress in tocopherol-deficient Arabidopsis mutants growing in a hydroponic system. Plant Signal Behav. 2013;8(2): e23136–e23136.

[82] Semida WM, Taha R, Abdelhamid MT, Radi MM. Foliar-applied α-tocopherol enhances salt-tolerance in *Vicia faba* L. plants grown under saline conditions. S Afr J Bot. 2014;95:24-31

[83] Semida WM, El-Mageed, TAA, Howladar SM, Radi MM. Foliar-applied α-tocopherol enhances salt-tolerance in onion plants by improving antioxidant defence system. Aust J Crop Sci. 2016;10(7):1030-1039.

[84] Goodwin TW. Nature and distribution of carotenoids. Food Chem. 1980;5(1):3-13.

[85] Ramos A, Coesel S, Marques A, Rodrigues M, Baumgartner A, Noronha J, et al. Isolation and characterization of a stress-inducible *Dunaliella salina* Lcy-β gene encoding a functional lycopene β-cyclase. Appl Microbiol Biotechnol. 2008;79(5): 819-28.

[86] Paliwal C, Mitra M, Bhayani K, Bharadwaj SVV, Ghosh T, Dubey S, et al. Abiotic stresses as tools for metabolites in microalgae. Bioresour Technol. 2017;244:1216-26.

[87] Ruiz-Sola MÁ, Rodríguez-Concepción M. Carotenoid biosynthesis in Arabidopsis: a colorful pathway. Arab B.2012;10:e0158–e0158.

[88] Jin, Chao, Ji, Jing, Zhao, Qing, Ma, Ran, Guan, Chunfeng, Wang, Gang. Characterization of lycopene β-cyclase gene from *Lycium chinense* conferring salt tolerance by increasing carotenoids synthesis and oxidative stress resistance in tobacco. Mol Breed. 2015; 35(12):228.

[89] Berera R, van Stokkum IHM, Kennis JTM, Grondelle R van, Dekker JP. The light-harvesting function of carotenoids in the cyanobacterial

stress-inducible IsiA complex. Chem Phys. 2010;373(1):65-70.

[90] Kang L, Ji CY, Kim SH, Ke Q, Park S-C, Kim HS, et al. Suppression of the β-carotene hydroxylase gene increases β-carotene content and tolerance to abiotic stress in transgenic sweetpotato plants. Plant Physiol Biochem. 2017;117:24-33.

[91] Li C, Ji J, Wang G, Li Z, Wang Y and Fan Y. Over-Expression of *LcPDS*, *LcZDS*, and *LcCRTISO*, genes from wolfberry for carotenoid biosynthesis, enhanced carotenoid accumulation, and salt tolerance in tobacco. Front Plant Sci. 2020; 11:119. doi: 10.3389/fpls.2020.00119

[92] Tuteja N. Mechanisms of high salinity tolerance in plants. Methods Enzymol. 2007;428:419-38.

[93] Flowers TJ, Colmer TD. Salinity tolerance in halophytes. New Phytol. 2008;179(4):945-63.

[94] Ghosh UK, Islam MN, Siddiqui MN, Khan MAR. Understanding the roles of osmolytes for acclimatizing plants to changing environment: a review of potential mechanism. Plant Signal Behav. 2021;16(8):1913306.

[95] Kalaji HM, Govindjee, Bosa K, Kościelniak J, Żuk-Gołaszewska K. Effects of salt stress on photosystem II efficiency and CO2 assimilation of two Syrian barley landraces. Environ Exp Bot. 2011;73:64-72.

[96] Ahn C, Park U, Park PB. Increased salt and drought tolerance by d-ononitol production in transgenic *Arabidopsis thaliana*. Biochem Biophys Res Commun. 2011;415(4):669-74.

[97] Fougère F, Le Rudulier D, Streeter JG. Effects of Salt stress on amino acid, organic acid, and carbohydrate composition of roots, bacteroids, and cytosol of alfalfa (*Medicago sativa* L.). Plant Physiol. 1991;96(4):1228-36.

[98] Zuther E, Koehl K, Kopka J. Comparative metabolome analysis of the salt response in breeding cultivars of rice. In: Jenks MA, Hasegawa PM, Jain SM. editors. Advances in Molecular Breeding Toward Drought and Salt Tolerant Crops. Dordrecht Springer; !st ed. 2007. p. 285-315

[99] Nounjan N, Nghia PT, Theerakulpisut P. Exogenous proline and trehalose promote recovery of rice seedlings from salt-stress and differentially modulate antioxidant enzymes and expression of related genes. J Plant Physiol. 2012;169(6):596-604.

[100] Wu G-Q, Feng R-J, Li S-J, Du Y-Y. Exogenous application of proline alleviates salt-induced toxicity in sainfoin seedlings. J Anim Plant Sci. 2017;27(1):246-251.

[101] Tarczynski MC, Jensen RG, Bohnert HJ. Expression of a bacterial mtlD gene in transgenic tobacco leads to production and accumulation of mannitol. Proc Natl Acad Sci U S A. 1992;89(7):2600-4.

[102] Kahlaoui B, Hachicha M, Misle E, Fidalgo F, Teixeira J. Physiological and biochemical responses to the exogenous application of proline of tomato plants irrigated with saline water. J Saudi Soc Agric Sci. 2018;17(1):17-23.

[103] Sabagh A El, Sorour S, Omar AE, Ragab A, Islam MS, Ueda A, et al. Alleviation of adverse effects of salt stress on soybean (*Glycine max* L.) by using osmoprotectants and organic nutrients. Int. J. Agric. Biol. 2015;9(9):1014-8.

[104] Shahid MA, Balal RM, Pervez MA, Abbas T, Aqeel MA, Javaid MM, et al. Exogenous proline and proline-enriched *Lolium perenne* leaf extract protects

against phytotoxic effects of nickel and salinity in *Pisum sativum* by altering polyamine metabolism in leaves. Turk J Bot. 2014; 38(5): 914-926.

[105] Zhang J, Zhang Y, Du Y, Chen S, Tang H. Dynamic metabonomic responses of tobacco (*Nicotiana tabacum*) plants to salt stress. J Proteome Res. 2011;10(4):1904-14.

[106] de Oliveira DF, Lopes L de S, Gomes-Filho E. Metabolic changes associated with differential salt tolerance in sorghum genotypes. Planta. 2020;252(3):34.

[107] Garg N, Sharma A. Role of putrescine (Put) in imparting salt tolerance through modulation of put metabolism, mycorrhizal and rhizobial symbioses in *Cajanus cajan* (L.) Millsp. Symbiosis. 2019;79(1):59-74. DOI:10.1007/s13199-019-00621-7

[108] Roy P, Niyogi K, SenGupta DN, Ghosh B. Spermidine treatment to rice seedlings recovers salinity stress-induced damage of plasma membrane and PM-bound H+-ATPase in salt-tolerant and salt-sensitive rice cultivars. Plant Sci. 2005;168(3):583 591. DOI:10.1016/j.plantsci.2004.08.014

[109] Roy M, Wu R. Arginine decarboxylase transgene expression and analysis of environmental stress tolerance in transgenic rice. Plant Sci. 2001;160(5):869-75.

[110] Harinasut P, Tsutsui K, Takabe T, Nomura M, Takabe T, Kishitani S. Exogenous glycinebetaine accumulation and increased salt-tolerance in rice seedlings. Biosci Biotechnol Biochem. 1996;60(2):366-8.

[111] Malekzadeh P. Influence of exogenous application of glycinebetaine on antioxidative system and growth of salt-stressed soybean seedlings (*Glycine max* L.). Physiol Mol Biol Plants. 2015;21(2):225-32.

[112] Abdelmotlb NA, Abdel-All FS, Abd EL-Hady SA, EL-Miniawy SM, Ghoname AA. Glycine betaine and sugar beet extract ameliorated salt stress adverse effect on green bean irrigated with saline water. Middle East J Appl Sci. 2019;9(1):142-154.

[113] Downton WJS. Photosynthesis in salt-stressed grapevines. Funct Plant Biol. 1977;4(2):183-92.

[114] Murakeözy ÉP, Nagy Z, Duhazé C, Bouchereau A, Tuba Z. Seasonal changes in the levels of compatible osmolytes in three halophytic species of inland saline vegetation in Hungary. J Plant Physiol. 2003;160(4):395-401.

[115] Garg AK, Kim J-K, Owens TG, Ranwala AP, Choi Y Do, Kochian L V, et al. Trehalose accumulation in rice plants confers high tolerance levels to different abiotic stresses. Proc Natl Acad Sci. 2002;99(25):15898-903.

[116] Nishizawa A, Yabuta Y, Shigeoka S. Galactinol and raffinose constitute a novel function to protect plants from oxidative damage. Plant Physiol. 2008;147(3):1251-63.

[117] Tarczynski MC, Jensen RG, Bohnert HJ. Expression of a bacterial mtlD gene in transgenic tobacco leads to production and accumulation of mannitol. Proc Natl Acad Sci USA. 1992;89(7):2600-4.

[118] Sheveleva E, Chmara W, Bohnert HJ, Jensen RG. Increased salt and drought tolerance by D-ononitol production in transgenic *Nicotiana tabacum* L. Plant Physiol. 1997;115(3):1211-9.

[119] Majee M, Maitra S, Dastidar KG, Pattnaik S, Chatterjee A, Hait NC, et al. A novel salt-tolerant L-myo-inositol-1-phosphate synthase from *Porteresia coarctata* (Roxb.) Tateoka, a halophytic wild rice: molecular cloning, bacterial overexpression, characterization, and

functional introgression into tobacco-conferring salt toleranc. J Biol Chem. 2004;279(27):28539-52.

[120] Hare PD, Cress WA. Metabolic implications of stress-induced proline accumulation in plants. Plant Growth Regul. 1997;21(2):79-102. DOI:10.1023/A:1005703923347

[121] Kumar R. Role of naturally occurring osmolytes in protein folding and stability. Arch Biochem Biophys. 2009;491(1-2):1-6.

[122] Sami F, Yusuf M, Faizan M, Faraz A, Hayat S. Role of sugars under abiotic stress. Plant Physiol Biochem. 2016;109:54-61.

[123] Cuin TA, Shabala S. Amino acids regulate salinity-induced potassium efflux in barley root epidermis. Planta. 2007;225(3):753-61.

[124] Butterworth PJ. Lehninger: principles of biochemistry (4th edn) D. L. Nelson and M. C. Cox, W. H. Freeman & Co., New York, 1119 pp. Cell Biochem Funct. 2005;23(4):293-4.

[125] Barnett NM, Naylor AW. Amino Acid and protein metabolism in bermuda grass during water stress. Plant Physiol. 1966;41(7):1222-30.

[126] Sanchez DH, Siahpoosh MR, Roessner U, Udvardi M, Kopka J. Plant metabolomics reveals conserved and divergent metabolic responses to salinity. Physiol Plant. 2008;132(2):209-19.

[127] Widodo, Patterson JH, Newbigin E, Tester M, Bacic A, Roessner U. Metabolic responses to salt stress of barley (*Hordeum vulgare* L.) cultivars, Sahara and Clipper, which differ in salinity tolerance. J Exp Bot. 2009;60(14):4089-103.

[128] Rare E. Stress physiology: The functional significance of the accumulation of nitrogen-containing compounds. J Hortic Sci. 1990;65(3):231-43. DOI:10.1080/00221589.1990.11516052

[129] Mansour MMF. Nitrogen containing compounds and adaptation of plants to salinity stress. Biol Plant. 2000;43(4):491-500. DOI:10.1023/A:1002873531707

[130] Mattioli R, Costantino P, Trovato M. Proline accumulation in plants: not only stress. Plant Signal Behav. 2009;4(11):1016-8.

[131] Szabados L, Savouré A. Proline: a multifunctional amino acid. Trends Plant Sci. 2010;15(2):89-97.

[132] Liang X, Zhang L, Natarajan SK, Becker DF. Proline mechanisms of stress survival. Antioxid Redox Signal. 2013;19(9):998-1011.

[133] Rai AN, Penna S. Molecular evolution of plant P5CS gene involved in proline biosynthesis. Mol Biol Rep. 2013;40(11):6429-35. DOI:10.1007/s11033-013-2757-2

[134] Hayat S, Hayat Q, Alyemeni MN, Wani AS, Pichtel J, Ahmad A. Role of proline under changing environments: a Review. Plant Signal Behav. 2012;7(11):1456-66.

[135] Shahbaz M, Ashraf M. Improving Salinity Tolerance in Cereals. CRC Crit Rev Plant Sci. 2013 Jul 4;32(4):237-49. DOI:10.1080/07352689.2013.758544

[136] Akçay N, Bor M, Karabudak T, Özdemir F, Türkan I. Contribution of gamma amino butyric acid (GABA) to salt stress responses of *Nicotiana sylvestris* CMSII mutant and wild type plants. J Plant Physiol. 2012;169(5):452-8.

[137] Bouché N, Fromm H. GABA in plants: just a metabolite? Trends Plant Sci. 2004;9(3):110-5.

[138] Song H, Xu X, Wang H, Wang H, Tao Y. Exogenous gamma-aminobutyric acid alleviates oxidative damage caused by aluminium and proton stresses on barley seedlings. J Sci Food Agric. 2010;90(9):1410-6.

[139] Renault H, Roussel V, El Amrani A, Arzel M, Renault D, Bouchereau A, et al. The Arabidopsis pop2-1 mutant reveals the involvement of GABA transaminase in salt stress tolerance. BMC Plant Biol. 2010;10:20.

[140] Alet AI, Sánchez DH, Cuevas JC, Marina M, Carrasco P, Altabella T, et al. New insights into the role of spermine in *Arabidopsis thaliana* under long-term salt stress. Plant Sci. 2012;182:94-100.

[141] Alcázar R, Marco F, Cuevas JC, Patron M, Ferrando A, Carrasco P, et al. Involvement of polyamines in plant response to abiotic stress. Biotechnol Lett. 2006;28(23):1867-76.

[142] Patel N, Gantait S, Panigrahi J. Extension of postharvest shelf-life in green bell pepper (*Capsicum annuum* L.) using exogenous application of polyamines (spermidine and putrescine). Food Chem. 2019;275: 681-7.

[143] Liu Y, Liang H, Lv X, Liu D, Wen X, Liao Y. Effect of polyamines on the grain filling of wheat under drought stress. Plant Physiol Biochem PPB. 2016; 100:113-29.

[144] Krasensky J, Jonak C. Drought, salt, and temperature stress-induced metabolic rearrangements and regulatory networks. J Exp Bot. 2012;63(4):1593-608.

[145] Robinson SP, Jones GP. Accumulation of glycinebetaine in chloroplasts provides osmotic adjustment during salt stress. Funct Plant Biol. 1986;13(5):659-68. DOI:10.1071/PP9860659

[146] Genard H, Saos JL, Billard J, Trémolières A, Boucaud J. Effect of salinity on lipid composition, glycine betaine content and photosynthetic activity in chloroplasts of *Suaeda maritima*. Plant Physiol Biochem. 1991;29:421-7.

[147] Ashraf M, Harris PJC. Potential biochemical indicators of salinity tolerance in plants. Plant Sci. 2004;166(1):3-16.

[148] Hanson AD, Scott NA. Betaine Synthesis from Radioactive Precursors in Attached, Water-stressed Barley Leaves. Plant Physiol. 1980;66(2):342-8.

[149] Mäkelä P, Jokinen K, Kontturi M, Peltonen-Sainio P, Pehu E, Somersalo S. Foliar application of glycinebetaine—a novel product from sugar beet—as an approach to increase tomato yield. Ind Crops Prod. 1998;7(2):139-48.

[150] Hossain MA, Hasanuzzaman M, Fujita M. Coordinate induction of antioxidant defense and glyoxalase system by exogenous proline and glycinebetaine is correlated with salt tolerance in mung bean. Front Agric China. 2011;5:1-14.

[151] Dubey RS, Singh AK. Salinity Induces accumulation of soluble sugars and alters the activity of sugar metabolising enzymes in rice Plants. Biol Plant. 1999;42(2):233-9. DOI:10.1023/A:1002160618700

[152] Cram WJ. Negative feedback regulation of transport in cells. The maintenance of turgor, volume and nutrient supply. In: Lüttge U, Pitman MG, editors. Transport in Plants II: Part A Cells. 1 st. ed. Heidelberg: Springer; 1976. p. 284-316. DOI:10.1007/978-3-642-66227-0_11

[153] Rathert G. The influence of high salt stress on starch, sucrose and degradative enzymes of two *glycine max*

varieties that differ in salt tolerance. J Plant Nutr. 1985;8(3):199-209.

[154] Crowe JH, Crowe LM, Chapman D. Preservation of membranes in anhydrobiotic organisms: the role of trehalose. Science. 1984;223(4637): 701-3.

[155] Vandesteene L, Ramon M, Le Roy K, Van Dijck P, Rolland F. A single active trehalose-6-P synthase (TPS) and a family of putative regulatory TPS-like proteins in Arabidopsis. Mol Plant. 2010;3(2):406-19.

[156] Habibur Rahman Pramanik M, Imai R. Functional identification of a trehalose 6-phosphate phosphatase gene that is involved in transient induction of trehalose biosynthesis during chilling stress in rice. Plant Mol Biol. 2005;58(6):751-62.

[157] Peterbauer T, Richter A. Biochemistry and physiology of raffinose family oligosaccharides and galactosyl cyclitols in seeds. Seed Sci Res. 2001;11(3):185-197.

[158] Jefferies RL. The Role of organic solutes in osmoregulation in halophytic higher plants. In: Rains DW, Valentine RC, Hollaender A, editors. Genetic Engineering of Osmoregulation: Impact on Plant Productivity for Food, Chemicals, and Energy. 1 st ed. Boston: Springer; 1980. p. 135-54.

[159] Smirnoff N, Cumbes QJ. Hydroxyl radical scavenging activity of compatible solutes. Phytochemistry. 1989;28(4):1057-60.

[160] Tari I, Kiss G, Deér AK, Csiszár J, Erdei L, Gallé Á, et al. Salicylic acid increased aldose reductase activity and sorbitol accumulation in tomato plants under salt stress. Biol Plant. 2010;54(4):677-83.

[161] Williamson JD, Jennings DB, Guo W-W, Pharr DM, Ehrenshaft M. Sugar alcohols, salt stress, and fungal resistance: polyols—Multifunctional Plant Protection? J Am Soc Hortic Sci. 2002; 127(4):467-73.

[162] Adams P, Thomas JC, Vernon DM, Bohnert HJ, Jensen RG. Distinct cellular and organismic responses to salt stress. Plant Cell Physiol. 1992;33(8):1215-23. DOI:10.1093/oxfordjournals.pcp.a078376

Chapter 2

Salt and Water Stress Responses in Plants

Mirela Irina Cordea and Orsolya Borsai

Abstract

Climate change-driven ecological disturbances have a great impact on freshwater availability which hampers agricultural production. Currently, drought and salinity are the two major abiotic stress factors responsible for the reduction of crop yields worldwide. Increasing soil salt concentration decreases plant water uptake leading to an apparent water limitation and later to the accumulation of toxic ions in various plant organs which negatively affect plant growth. Plants are autotrophic organisms that function with simple inorganic molecules, but the underlying pathways of defense mechanisms are much more complex and harder to unravel. However, the most promising strategy to achieve sustainable agriculture and to meet the future global food demand, is the enhancement of crop stress tolerance through traditional breeding techniques and genetic engineering. Therefore, it is very important to better understand the tolerance mechanisms of the plants, including signaling pathways, biochemical and physiological responses. Although, these mechanisms are based on a well-defined set of basic responses, they can vary among different plant species.

Keywords: abiotic stress, salinity, drought, response mechanisms, tolerance

1. Introduction

Salinity and drought are the two major constraints that affect plant growth and crop production alongside other stress conditions such as extreme temperature, heavy metals, flooding etc. thus reducing agricultural productivity worldwide. Both the cellular and molecular responses of plants to these environmental stresses have already been investigated, however understanding these mechanisms by which plants can perceive stress signals and transmit them to cellular machinery to activate adaptive responses is a very important chain-link of plant physiology. Besides, extending knowledge about stress signal transduction becomes vital for breeding programs and genetic engineering to improve stress tolerance in crops.

Due to climate change, it is predicted that drought and salinity will became more severe in the upcoming years which could lead to a significant reduction of plant growth and yield of several economically important species. It has been estimated that worldwide food demand will increase by 70% until the end of 2050 [1] due to a population growth of 2.3 billion people. In this context, developing crop plants with high yield and better tolerance to harsh environmental conditions becomes an urgent need to meet future food demand for next generations.

In general, plant responses to salinity and drought may vary in morphological, physiological and biochemical aspects and processes. Most of the effects induced by salinity and drought are negative, however to some extent they can have positive effects as well [2]. It has been reported that salinity at certain concentrations enhanced plant fecundity due to an increase in reproduction, but it has also been observed that this enhancement was highly dependent on genotype and plant developmental stage [3]. Soil water salinity can also have a positive effect on fine particles helping them to bind together into aggregates, thus improving soil aeration, root penetration and root growth [4]. Nevertheless, salinity cannot be increased in favor of soil structure without considering the potential impacts on plant health.

Salt-stress resistance represents the ability of a plant to prevent, reduce or overcome the possible damaging effects caused directly or indirectly by the presence of excessive soluble salts (accumulation of toxic ions) in its root zone. A 50% reduction in yield can be considered a measure of salt stress.

Drought stress occurs after a relatively long period with no rains, inducing moisture stress in the soil detrimental to crop growth, especially in rainfed agriculture. The severity of drought is strongly related to the timing (growth stage of the plants) and intensity (duration of no rain period). Other factors such as soil characteristics and agricultural practices can interfere with crop yields.

Previous reports suggest that a positive transgenerational impact on seedling vigor of *Brassica napus* has been observed due to drought stress [5]. This phenomenon was explained as a result of the heterotic effects, altered reservoir of seed storage metabolites, and inter-generational stress memory formed by stress-induced changes in the epigenome of the seedling. Compared to salt stress, drought stress has more severe effects on plants and economy [6] but plant responses are closely related and their defense mechanisms even overlap.

The ability of a crop variety to perform better over other varieties under drought conditions is known as drought resistance which is linked to achieved yields and potential yields achievable in a given environment in the absence of drought conditions. Drought resistance is highly environment specific and yield stability might be influenced by crop management practices, and/or physiological mechanisms and might not necessarily be associated with the drought resistance ability of a genotype. In a drought resistant variety, plant growth and development are well-matched to specific drought environment(s) [7].

When sensing salinity or drought stresses, plants have the capability to combine a range of responses in order to avoid stress injuries and complete their life cycle. By the activation of various defense mechanisms plants can store reserves in their organs and use them later for yield production or, they can tolerate stress conditions without tissue dehydration [8]. Plant-associated organisms play an important role in improving the adaptation strategies of plants to environmental stresses. In this context, microorganisms, for example, can rescue plants from the deleterious effects of drought and salinity through their activity, such as nutrient solubilization, IST and production of phytohormones (IAA, Cytokinin, ABA or GA), EPS and ACC deaminase. The inoculation of plants with arbuscular mycorrhizal fungus can also increase plants' tolerance to short term salinity exposures [9, 10].

With all these fundamentals being provided to understand the underlying defense mechanisms of plants against stress conditions, further studies are still needed to reveal key mechanisms which govern salinity and drought tolerance responses in plants and which can lead us towards better direction in crop improvement, in order to obtain potential candidates for future saline agriculture.

2. Mechanism of salt stress and plant response

Stress factors, such as osmotic, ion toxicity, nutrient imbalance or soil pH alter the expression of several morphological, physiological and biochemical characteristics of plants. As the stress increases, plant growth is further restricted. Under severe stress conditions plants may die prematurely after germination or transplanting or can survive longer shriveling [11, 12].

Seed germination is often hindered and/or delayed when environmental stresses occur. Seedlings often fail to survive since in this stage of growth plants are the most vulnerable [13]. Plant growth is stunted affecting most of the vegetative characters, such as leaf number, size, shoot number, plant height etc. [14, 15]. Regarding the reproductive traits of the plants, salt stress can often induce an early flowering and abortion of flower buds [16, 17]. Furthermore, a significant overall reduction in yield can be observed in most of the plant species subjected to salt stress. Achieved yields are usually much lower than potential yields under normal growing conditions [18–20].

Plant growth in saline soils is usually affected because of the osmotic effect in the soil solution. High salt concentration increases the potential forces that hold water in the soil and makes it more difficult for plant roots to extract soil moisture. During dry periods, salt in soil solutions may be so concentrated as to kill plants by sucking water from them (exosmosis) [21]. Moreover, salt in the soil solution forces a plant to exert more energy to absorb water and to exclude salt from metabolically active sites. As salinity increases, plant growth is further restricted. A saline soil should be kept wet to dilute the salt concentration so as to cause the least salt hindrance to the growing plants. Also, plant growth in sodic/alkaline soils is affected due to high ESP throughout the profile, very low infiltration and hydraulic conductivity rates [22]. The exchangeable complex of alkaline soils is largely occupied by sodium ions which cause dispersion of soil due to the breakdown of aggregates forming a dense surface crust which greatly hinders seedling emergence due to low permeability of the soil to water and air. Poor drainage in such soils is due to a high water table which further restricts plant's ability to absorb water and nutrients in required amounts [23]. High pH results in reduced availability of some essential plant nutrients [24]. Accumulation of certain elements in plant parts at toxic levels may result in plant injury or reduced growth and even death in extreme cases. The most common toxic elements are sodium, molybdenum and boron. Selenium may also occur in toxic concentrations. Plant growth in degraded alkaline or solodic soils is largely due to poor drainage.

Crop species and varieties greatly vary regarding their response to salt stress (**Figures 1** and **2**). Many naturally occurring plants in salt-affected soils (halophytes) have certain specific structures and adaptation strategies, for example salt glands and salt hairs on their leaves [25, 26]. Detailed studies on salt glands in salt-tolerant plants, such as the halophyte kallar grass, *Leptochloa fusca*, showed the presence of enlarged cells protruding above the epidermis of both abaxial and adaxial surfaces of leaves and also on the exposed side of the leaf sheath [27]. These glands are associated with salt deposition (Na > K > Ca > Mg) on leaf surfaces. *Acanthus ilicifolius* and other crop species have salt glands on the adaxial leaf surface and studies have shown each gland to be surrounded by six collecting cells (salt-collecting cells) [28]. One of the most salt-tolerant plants, the halophytic wild rice, *Porteresia coarctata* has unicellular salt hairs on the adaxial surface of the leaves. Analysis of its leaf washing showed that Na and Cl were predominantly excreted, followed by K, Mg and Ca [29]. In other species such as *Puccinellia tenuifolia* the phenomenon of salt excretion has also been observed [30]. Moreover, some crop species have sunken stomata associated with the occurrence of high density of trichomes arising from the epidermis, as an adaptive mechanism to minimize water loss under stressful habitats [31].

Plants subjected to salt stress face the problem of reduced availability of water and response to changes in the processes related to maintenance of a favorable water balance [32, 33]. According to previous reports, the increase in salinity

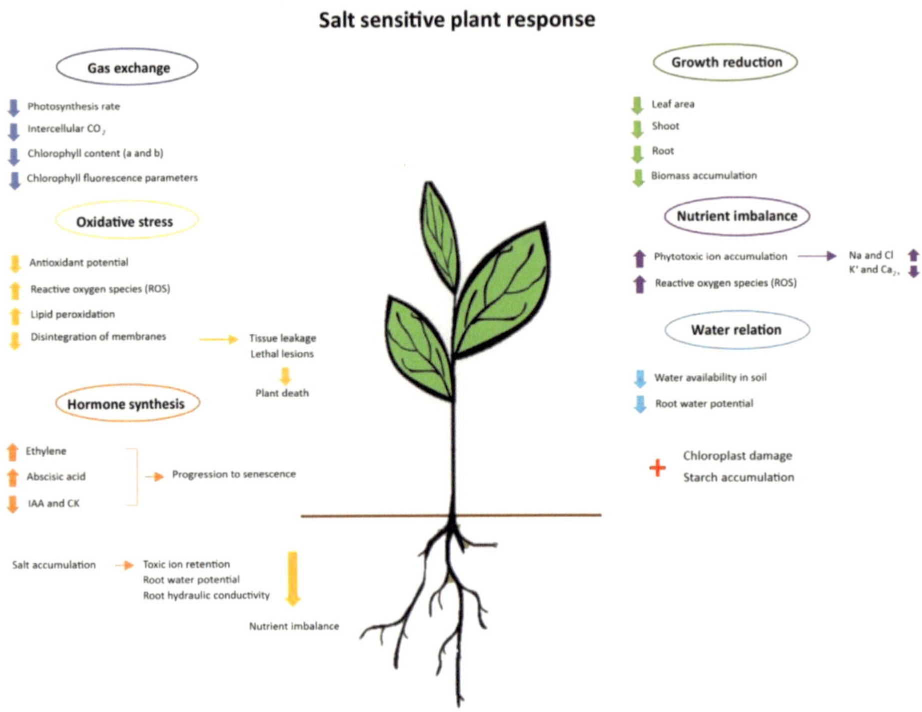

Figure 1.
The effect of salinity on salt-sensitive plants.

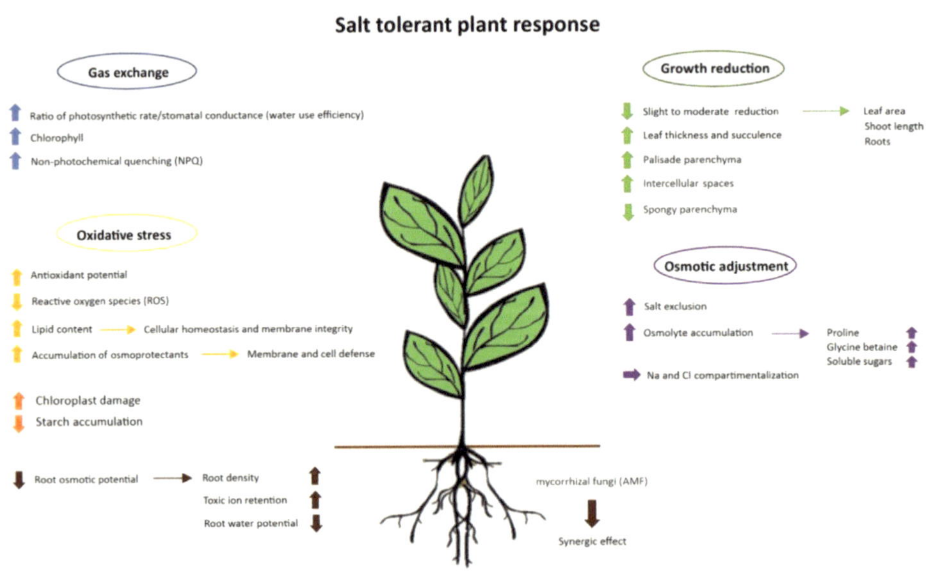

Figure 2.
The effect of salinity on salt-tolerant plants.

resulted in a decrease in transpiration in mustard [34], quinoa [35], wheat and pearl millet [36, 37], whereas leaf diffusive resistance (LDR) and leaf temperature increased. Higher LDR coupled with low transpiration might contribute to moisture conservation in plants under salt stress conditions [38].

Excessive salt in the root zone not only reduces the availability of water to plants, but their excessive absorption of salt increases the risk of ion toxicity and interference in the uptake of other essential nutrients [39]. Several reports indicate that increasing salinity and sodicity (Na content) decreases K ion concentration [40–42]. The antagonistic effect of both cations is well established. Tolerant varieties show a tendency to take up less Na while maintaining their K status.

Furthermore, plants growing at sublethal levels of salt stress may often appear greener due to increase in chlorophyll [43, 44]. Accumulation of certain amino acids, sugars and other osmotically active organic substances in response to salt stress are indications of altered nitrogen and carbohydrate metabolism. In this regard, it has been observed, for example, that two-week-old wheat plants doubled their amino acid content after 24 hours when subjected to electrolyte concentration (EC) of 22. Amino acids are very important components of plants, exhibiting various roles. Under abiotic stress conditions they can act as osmolytes, regulate the ion transport in the plant or regulate the stomatal opening and closure [45]. Besides, they can contribute to diverse enzyme synthesis improving plant abiotic stress tolerance through gene expression [46]. Among amino acids, glutamine (Glu), phenylalanine (Phe) and proline (Pro) proved to have significant roles in response to salt stress condition such as signaling precursors (Glu), building blocks of plant structure (Phe) and beneficial solutes (Pro). In this regard, previous research results show a considerable increase in glutamine, phenylamine and especially in proline content as a response to salt stress improving plant tolerance or indicating its sensitivity [39]. In general, the highest proline accumulation occurs in lamina followed by leaf sheath, stems or shoots and roots as observed in several plant species such as *Phaseolus sp.*, *Portulaca sp.*, *Triticum sp.*, *Solanum lycopersicum* etc. (**Table 1**) [57–61]. Moderately tolerant barley varieties accumulated more proline than sensitive ones [62].

In wheat, water-soluble proteins increased in leaves in response to salinity [63]. Another example, such as rhodes grass, *Chloris gayana*, could be given for the increase of trichloroacetic acid and NaOH soluble proteins in response to salinity [64]. Enzymes are also influenced by change in plant water status as well as ionic imbalance [65, 66]. Decrease in (a) amylase activity with increase in salinity was observed in wheat and chickpea leaves after short term exposure to salt stress while activity of invertase and other enzymes of carbohydrate metabolism significantly increased [67, 68]. Nitrate reductase activity may also decrease with increase in stress level in many species [69, 70]. Tolerant varieties of pearl millet showed a tendency to maintain their nitrate-reductase activity [71]. Polyphenol oxidase activity has been reported to be higher in sensitive varieties of wheat, barley and rice [72–74].

Due to their occasional or constants exposure to harsh, unfavorable environmental conditions, plants developed a series of detoxification mechanisms to be able to maintain their growth and alleviate potential damages caused by 'reactive oxygen species' (ROS) - at cellular level [75].

Oxidative damage in plants often occurs as a secondary effect of different harmful environmental conditions such as drought, salinity, cold, heat, or heavy metals in the soil. Under these conditions, the level of ROS can largely increase overwhelming plant defense systems, and thus inducing multiple deleterious effects at the cellular level. These effects are the result of the oxidation of membrane lipids, amino acid residues in proteins and the bases in DNA. In general, plants respond to an increase in ROS by activating enzymatic or non-enzymatic antioxidant processes to overcome ROS accumulation. Among them, malondialdehyde (MDA), a lipid peroxidation product is considered a reliable oxidative stress marker not only in

Plant species	Amino acids	Increase of amino acids	Salt concentration (NaCl)	References
Triticum aestivum	Glutamine	1.33-fold 2.02-fold	150 mM 300 mM	[47]
Anacardium occidentale	Glutamine	1.37-fold	100 mM	[48]
Oryza sativa L. cv. Kinuhikari	Glutamine	1.5-fold	150 mM	[49]
Helianthus annuus L. cv. SH222	Glutamine	6.2-fold	126 mM	[50]
Jatropha curcas L.	Phenylalanine	1.12-fold	150 mM	[51]
Salvia sp.	Phenylalanine	12–18-fold	100 mM	[52]
Solanum nigrum	Phenylalanine	23-fold	150 mM	[53]
Zea mays L.	Phenylalanine	2.26-fold	150 mM	[54]
T. aestivum	Proline	2.26-fold 19.29-fold	150 mM 300 mM	[62]
Solanum tuberosum L.	Proline	3.4-fold	250 mM	[54]
Hordeum vulgare	Proline	20–31-fold	300 mM	[55]
A. occidentale	Proline	22-fold	100 mM	[48]
Solanum lycopersicum L.	Proline	3-fold	60 mM	[61]
Portulaca halimoides	Proline	5.66-fold	400 mM	[59]
Phaseolus vulgaris L.	Proline	2.6-fold	150 mM	[56]

Table 1.
Prominent amino acids and their changes in responses to salt stress.

plants but in animals also, which is generated by the oxidation of membrane lipids [76]. Several scientific reports show an increase of MDA levels in response to abiotic stresses in various plant species: rice, *Calendula*, *Miscanthus*, basil, *Solanum* and many others [77–81].

Moreover, phenolic compounds are known to have multiple roles in plants; some of them being part of the structural component of cell walls, while others are involved in growth regulation and developmental processes or the activation of defense mechanisms against biotic and abiotic stresses. Several reports also describe the mediatizing effects of antioxidant properties of many phenolic compounds on plant responses to salinity and drought showing an increase in their content under high salinity and water deficit conditions [82, 83].

Flavonoids, the most complex subclass of phenolic compounds are also involved in a wide-range of environmental interactions. The biosynthesis of flavonoids in plants is upregulated not only by UV-radiation but also in response to diverse biotic and abiotic stresses, from the depletion of mineral nutrients to salinity, cold or drought [84]. Previous studies suggest that flavonoid contents increase in plants when subjected to abiotic stress conditions and the accumulation of these compounds is tightly coupled with the intensity of the applied stress [85–87].

Ascorbic acid (Vitamin C) is one of the most powerful, water-soluble antioxidants as a scavenger ROS produced by most eukaryotic organisms. It occurs in all plant tissues, but mostly in the chloroplast, in mature leaves where these are fully developed and the chlorophyll levels are also the highest. It is considered the most important ROS detoxifying compound due to its ability to donate electrons in a number of enzymatic and non-enzymatic reactions [88].

Beside the above-mentioned compounds, α-tocopherols (vitamin E) are another family of antioxidants that can be found in all parts of the plants. They are the most biologically active and predominant antioxidants in the chloroplast membranes, and are mainly responsible for its protection against oxidative damages [89].

Antioxidant enzymes such as superoxide dismutase (SOD), several peroxidases (POD), catalase (CAT) and glutathione reductase (GR) play a crucial role as ROS scavengers in defense mechanisms against abiotic stresses. They are responsible for the maintenance of the proper redox equilibrium in plant cells [90]. Enzymatic activities have been studied in different plant species including both crop species and ornamental plants [91–93]. The results revealed that water stress, in general, led to a continuous increase of several antioxidant enzyme activities. In maize, for example, significant enhancements in the activities of several antioxidant enzymes (superoxide dismutase-SOD, catalase-CAT, ascorbate peroxidase-APX, and glutathione reductase-GR) occurred after 12 h of treatment showing an increase of 21%, 52% and 33% and 38% as compared to the control. It was also noticed that after 24 h of water stress treatment, the activities of the antioxidant enzymes showed a tendency to decrease when compared to the 12 h treatment [94].

3. Mechanism of drought resistance

Over the centuries plants have been exposed to different environmental conditions and applied diverse adaptation strategies to be able to cope with these challenges. Water deficit in plants occurs when the transpiration rate exceeds water uptake. Such water deficit is usual in most plants as a component of some developmental processes [95], but cellular water deficit can cause harmful changes in cell volume and membrane shape, disruption of water potential, decreased turgor pressure, or disruption of membranes. A total loss of free water will result in dehydration and plant loss. Plant responses to water deficit (**Figure 3**) primarily depend on the species and genotype, but also on the length and quantity of water loss, and the age and developmental stage of the plants. Among the complex plant mechanisms and regulatory networks for drought, osmotic adjustment plays an important role in water deficit avoidance, by lowering the water potential of the cells to support water uptake and maintain turgor. At molecular level, the accumulation of mRNA during

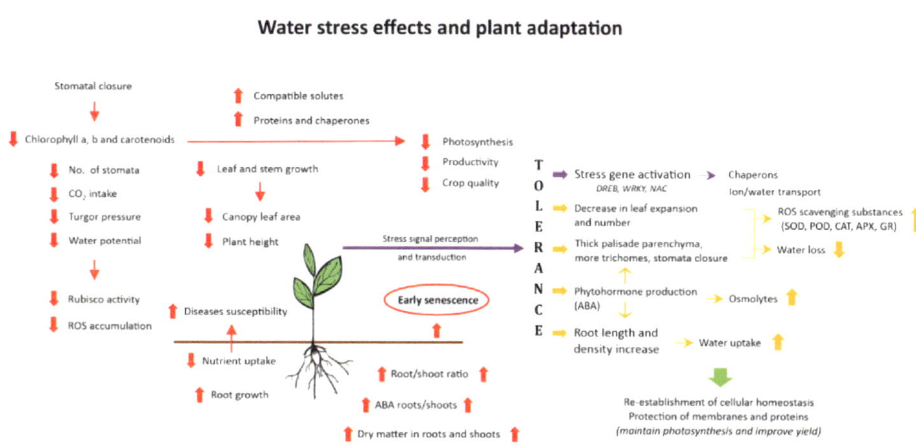

Figure 3.
Schematic representation of water stress effects and plant adaptation.

water deficit may indicate gene induction, but in order to obtain a fully functional gene product, other additional mechanisms such as translational regulation and posttranslational modification may be required. In general, plants respond to water deficit by employing some basic mechanisms to avoid water loss, protect the cellular machinery and repair damage [96, 97].

Susceptibility to drought can occur during the early vegetative seedling stage, during the period of panicle development prior to flowering, or/and during the post flowering stage of grain development [97]. Susceptibility during post-flowering stage is characterized by reduced seed size and grain yield, pre-mature plant and leaf senescence and increased stalk lodging [98]. Terminal post flowering drought results in an abbreviated period of grain development and therefore reduces seed size [97, 99]. Genotypes with a high rate and reduced duration of grain filling may be more tolerant under terminal post flowering conditions [100].

Identification of critical stages of crop growth, those at which a crop is more severely affected by drought and more particularly its response to stress, if any, is important to be known to be able to understand the mechanism of drought resistance. This knowledge could further help to develop appropriate methodology for developing drought-resistant varieties. The usual mechanisms are as follows:

1. *Drought escape*: is a strategy applied by plants in early maturing crops/crop varieties to complete the critical stages of crop growth before severe deficit occurs, focusing more on flowering and reproduction instead of developing new shoots and increasing leaf area [101]. Early growth vigor may enable a variety to establish a good plant stand rather quickly while the moisture supply is suitable. Thus, crops or crop varieties applying this strategy can escape the adverse effects of drought and perform relatively better. Many indeterminate crops respond to reirrigation by resuming their growth and still perform better [102].

2. *Avoidance*: Drought avoidance is an alternate mechanism by which plants can maintain positive tissue water relations even under limited soil moisture conditions. Mechanisms of drought avoidance typically involve water conservation at the whole plant level. Avoidance is accomplished by decreasing water loss from the shoot or by more efficiently extracting moisture from the soil [103]. Many crop varieties/crops with deep as well as dense root system may be able to maintain minimal water uptake from soil to avoid internal stress, at least during the initial stages [104]. High varietal resistance to water loss has also been observed in a few cases, for example, in wheat, rice the amount of epicuticular wax deposition is reportedly associated with water loss [105, 106]. Previous findings suggest that different species such as *Catharanthus roseus, Sorghum sp.* and *Oryza sativa* reduced transpiration rate by as much as 44 to 82% due to water stress [107–109].

3. *Tolerance:* Drought tolerance is defined in a number of ways, namely, the performance per se, the stability of performance under drought and last but not least specific physiological or morphological traits that are believed to be associated with the expression of drought tolerance. The mechanisms responsible for drought tolerance are functioning at tissue or cellular level [99]. When the tissue desiccates, these mechanisms are activated to stabilize and protect the cellular and metabolic integrity of the plant. Crop varieties may differ in their ability to thrive under drought conditions. This has been demonstrated through various test regarding physio-morphological and biochemical traits including desiccation survival, heat tolerance, osmolytes, ion homeostasis etc. [110–115].

4. *Recovery*: Drought stress conditions may vary in duration, but when rainfall does commence the ability of a genotype (or crop variety) to recover quickly and resume active growth is an important character. In rice, recovery capacity from drought is strongly related with characters such as vegetative growth vigor, high tillering ability, shallow root system and rather long growth duration [116]. Similar characters have been observed in different annual and perennial species, in wheat, sugarcane etc. [117–119]

3.1 Assessment of drought resistance and plant traits associated with drought resistance

Drought resistance of an annual crop plant can at present be assessed for agronomic purposes only on the basis of yield [120]. Few of the many screening tests proposed have been adopted by breeders.

Several plant traits, such as dehydration avoidance and dehydration tolerance have been found to be positively associated with yield under stress across genotypes of wheat and barley [121]. Leaf rolling, root system, pubescence of aerial organs, reflectance of incoming solar radiation, increased heat dissipation through decreased boundary layer resistance at the organ level (narrow leaves, awns), etc., are the main traits that contribute to dehydration avoidance. In nature, a better balance is associated with a higher proportion of energy dissipated as latent heat and hence a lower canopy temperature. Dehydration tolerance related to cellular and subcellular processes can be readily assessed by measurements of membrane stability with the electrolyte leakage test [122]. It is difficult, however, to relate this type of test to plant production. Nevertheless, visual scores on morphological traits, such as leaf rolling, root habit, etc., and/or observations recorded through other methods, if any, in relation to the above-mentioned characters should invariably be used as an indirect measurement of drought resistance for practicing selection in a breeding programme.

In sorghum, the 'stay-green' character is reportedly associated with post-flowering drought tolerance. Stay-green is characterized as resistance to premature leaf and stalk death induced by post-flowering drought. Resistance to premature leaf and stalk death is thought to increase the potential period of grain development and thereby stabilizing the expression of seed weight [123]. Sorghum lines with high levels of stay-green have been identified and are being used in some breeding programs [124–126].

3.2 Genetics of plant traits associated with drought resistance

A variety of adaptive plant characteristics related to environmental stress have been investigated and were shown to exhibit genetic variation. The variability of traits extends to the physiological, morphological and chemical characteristics of the plants. These three groups of traits are the most representative and useful markers for stress tolerance identification. Drought stress can cause many changes in the physiological traits, affecting the capability of plants to maintain high level of leaf-water potential under water deficit conditions, the osmotic adjustment and last but not least the capability of plants to recover after short or long-term rehydration. The regulation of photosynthesis, by stomatal closure and the stability of cellular membranes and its maintenance are crucial for plants to tolerate stress conditions. Osmolytes, such as Pro, glycine betaine and soluble sugars also play an important role in osmotic adjustment under various stress conditions, where accumulation may greatly vary among species. Morphological or phenotypic characters are considered important in the adaptation of plant to stress conditions, their responses being reflected and becoming quantifiable through root growth and density, leaf number size and canopy area, leaf orientation, stem or shoot length and number, flower development (number and fertility, seedling survival or any other trait specific for every species (leaf succulence, pubescence etc.) [127–133].

'Stay-green' or the capacity of green color retaining for longer time of the leaves after flowering is a desirable attribute for crop production. Sorghum genetic studies of 'stay-green' have generally indicated a complex pattern of inheritance. It has been reported that both dominant and recessive expression were strongly influenced by the environment. Previous reports reveled the inheritance of stay-green in a set of recombinant inbred lines of sorghum [134]. Due to a quantitative trait loci (QTL) mapping in sorghum for the extension of photosynthetic period 13 regions of the genome were identified and associated with the stay-green phenotype of post-flowering drought adaptation [135]. Two QTLs were successfully identified as the ones influencing yield and 'stay-green' capacity under post-flowering drought conditions. The same loci were also linked to yield under successful irrigation conditions indicating the pleiotropic nature of these tolerance loci on yield under favorable environmental conditions [136]. Similarly, the QTL mapping results suggested many other loci that were linked to the rate and duration of yield development [137]. The findings also revealed that high yield and short grain development were associated with instability of yield performance under water paucity [138].

It may be noted that associations between markers and QTL were somewhat variable across testing environments. This highlights the importance of multi-environment testing when evaluating drought tolerance.

Similar studies have been carried out in maze, where 15 green-leaf-area related QTLs were detected thus identifying the most important genomic region responsible for maintaining green leaf area at the final developmental stage of maize [139].

However, the current screening and breeding techniques allow to explore the genetic basis for various plants and identify diverse traits which help the plants to perform under stress conditions, high yield performance, good quality and stress resistance remains the eternal flame for crop breeders. These desirable crop production traits and their transmission from one genotype to another will remain attractive and unexplored [140].

In this regard, selection for drought and salt resistance will therefore continue to be primarily based on yield assessment under stress conditions [141].

4. Selection and breeding for salt and drought resistant varieties

Salt tolerance thresholds are usually set based on the relative crop yield at defined stress levels of salt stress. Besides, the biological traits of the plant are also of a great importance in the selection process since, these characters are the summary of genetic and environmental effects upon plant growth as a result of physiological processes, effects which confer salinity tolerance. Therefore, two primary selection criteria can be established for plant selections follow:

1. Seed germination capacity and seedling survival: Seed germination and seedling development, are the very early stages of plant development which are critical. Therefore, plants that can cope with salt stress conditions in these stages of their life cycle should be the prime requisite in the selection process for salt tolerance. Various crops and genotypes that even fail to establish themselves under defined stress conditions cannot be expected to do any better at a later stage of their growth.

2. Yield: Varieties highly tolerant to salinity are those that exhibit minimum reduction in relative economic yield with per unit increase in stress. The slope of regression of yield against stress gives a fairly reliable estimate of salt tolerance of a crop/genotype. This is by far the best index for identification and screening of salt-tolerant genotypes.

A number of other plant attributes, namely Na and K content in shoots/leaves, Na/K ratio, pH of the cell sap, proline content and enzyme response may also have some potential use. The only limitation to their practical use so far however, is, that the differential genotypic response observed in various crops cannot always be explained on the basis of these data. For this reason, the use of physiological characters is highly recommended to obtain more reliable information and select potential candidates for future saline agriculture.

The first step that should be taken to develop drought and salt resistant varieties is to identify drought-resistance QTLs, which are essential to set valuable candidates for crop breeding. Regarding the selection criteria, there are several promising traits to be targeted in breeding programmes as follows:

1. Root architecture – which plays an important role in drought avoidance of crops. Transcriptomic differences between deep and shallow rooting systems strongly influences the ATP synthesis. Such traits can significantly improve abiotic stress resistance in crops by introducing or manipulating a single gene;

2. ABA-synthesis which can improve drought resistance even at seedling stage in different crops;

3. Direct-deep-seeding tolerance of different species which could significantly contribute to water saving and drought resistance, for example in rice production;

4. Yield capacity under stress conditions;

5. Exploitation or domestication of wild relatives (halophytes) of crop plants. Interspecific hybridization has an important role in the improvement of crop plant performance under abiotic stress conditions.

In the evaluation process for plant tolerance to salt and drought stress, it is important to take into consideration all the three groups of traits (physiological, morphological and chemical characters) and evaluate plant responses as a whole. Due to great genetic variation of the plants, in some cases it is not enough to solely analyze the physiological, chemical or morphological profile since they are interconnected.

5. Conclusions

Recently, several research have been carried out to depict the complex underlying mechanisms (physiological, morphological and chemical) that control abiotic stress responses in crop plants. However, the exact genes, and their activation, which control plant defense mechanisms are still unclear. Tolerance against abiotic stresses in different crop plants has been improved by the application of transgenic technology of reactive oxygen species components, but future research studies are still needed to determine and increase yield performance and quality under harsh environmental conditions. Genetic improvement of crops needs to identify further genetic variations that allow plants to increase their tolerance against the upcoming abiotic stress levels than the ones we are facing today. It has to employ new tools to analyze the genetic, physiological and molecular basis of stress tolerance and to identify genes associated with improved resistance and integrate them into practical breeding to develop “smart” crop varieties which require lower input and provide high yield.

Conflict of interest

The authors declare no conflict of interest.

Author details

Mirela Irina Cordea and Orsolya Borsai*
University of Agricultural Sciences and Veterinary Medicine Cluj-Napoca, Cluj-Napoca, Romania

*Address all correspondence to: orsolya.borsai@usamvcluj.ro

References

[1] FAO statistical yearbook. Food and Agriculture Organization of the United Nations, Rome, 2013.

[2] Raza A, Razzaq A, Mehmood SS, Zou X, Zhang X, Lv Y, et al. Impact of climate change on crops adaptation and strategies to tackle its outcome: a review. Plants. 2019;8:34. DOI:10.104 6/j.1365-2745.2003.00806.

[3] Van Zandt PA, Tobler MA, Mouton E, Hasenstein KH, Mopper S. Positive and negative consequences of salinity stress for the growth and reproduction of the clonal plant, Iris hexagonal. Journal of Ecology. 2003;91:837-846. DOI: 10.1046/ j.1365-2745.2003.00806.x

[4] Igboama WN, Ugwu NU. Laboratory Evaluation of the Impact of Contaminants on Soil Resistivity and the Consequent Effect on Plant's Growth. Journal of Environmental Protection. 2016;7:12:1802.

[5] Hatzig SV, Nuppenau JN, Snowdon RJ, Schiessl S V. Drought stress has transgenerational effects on seeds and seedlings in winter oilseed rape (*Brassica napus* L.). BMC Plant Biology. 2018;18:297. DOI: 10.1186/s12870-018-1531-y

[6] Ma Y, Dias MC, Freitas H. Drought and salinity stress responses and microbe-induced tolerance in plants. Frontiers in Plant Science. 2020; 11:1750.

[7] Bacher H, Zhu F, Gao T, Liu K, Dhatt B K, Awada T, Walia H. Wild emmer introgression alters root-to-shoot growth dynamics in durum wheat in response to water stress. Plant Physiology. 2021. DOI:10.1093/ plphys/kiab292

[8] Mantri N, Patade V, Penna S, Ford R, Pang E. Abiotic Stress Responses in Plants: Present and Future. In: Ahmad P., Prasad M. editors. Abiotic Stress Responses in Plants. New York Springer, 2021. DOI:10.1007/ 978-1-4614-0634-1_1

[9] Scagel Carolyn F, Jungmin Lee. Salinity sensitivity and mycorrhizal responsiveness of polyphenolics in 'Siam Queen' basil grown in soilless substrate. Scientia Horticulturae. 2020;269: 109394.

[10] Pedranzani H, Rodríguezrivera M, Gutiérrez M, Porcel R, Hause B, Ruizlozano JM. Arbuscular mycorrhizal symbiosis regulates physiology and performance of *Digitaria eriantha* plants subjected to abiotic stresses by modulating antioxidant and jasmonate levels. Mycorrhiza. 2016;26:141-152. DOI:10.1007/s00572-015-0653-4.

[11] Yadav S, Modi P, Dave A, Vijapura A, Patel D, Patel M. Effect of abiotic stress on crops. Sustainable Crop Production. 2020.

[12] Lauer J. What happens within the corn plant when drought occurs. Corn Agronomis. 2003;10:22:153-155.

[13] Uçarlı C. Effects of Salinity on Seed Germination and Early Seedling Stage. In Abiotic Stress in Plants. IntechOpen; 2020.

[14] Zahra N, Raza ZA, Mahmood S. Effect of salinity stress on various growth and physiological attributes of two contrasting maize genotypes. Brazilian Archives of Biology and Technology. 2020;63.

[15] Isayenkov SV, Maathuis FJ. Plant salinity stress: many unanswered questions remain. Frontiers in Plant Science. 2019;10:80.

[16] Yuan F, Guo J, Shabala S, Wang B. Reproductive Physiology of Halophytes:

Current Standing. Frontiers in Plant Science. 2019;9:1954. DOI: 10.3389/fpls.2018.01954

[17] Vadez V, Rashmi M, Sindhu K, Muralidharan M, Pushpavalli R, Turner NC, ... Colmer TD. Large number of flowers and tertiary branches, and higher reproductive success increase yields under salt stress in chickpea. European Journal of Agronomy. 2012;41:42-51.

[18] Rao ES, Kadirvel P, Symonds RC, Ebert AW. Relationship between survival and yield related traits in *Solanum pimpinellifolium* under salt stress. Euphytica. 2013;190:2:215-228.

[19] He Y, Chen Y, Yu CL, Lu KX, Jiang QS, Fu JL, Wang GM, Jiang DA. Photosynthesis and yield traits in different soybean lines in response to salt stress. Photosynthetica. 2016;54:4:630-5.

[20] Di Caterina R, Giuliani MM, Rotunno T, De Caro A, Flagella Z. Influence of salt stress on seed yield and oil quality of two sunflower hybrids. Annals of Applied Biology. 2007;151:2:145-154.

[21] Augie MA, Adegbite MA. Salinity and Sodicity Status of the Floodplain Soils of Kebbi State, North Western, Nigeria. Asian Journal of Soil Science and Plant Nutrition. 2021;23-31.

[22] Crescimanno G, Iovino M, Provenzano G. Influence of salinity and sodicity on soil structural and hydraulic characteristics. Soil Science Society of America Journal. 1995;59:6:1701-1708.

[23] Xu X, Huang G, Sun C, Pereira LS, Ramos TB, Huang Q, Hao Y. Assessing the effects of water table depth on water use, soil salinity and wheat yield: Searching for a target depth for irrigated areas in the upper Yellow River basin. Agricultural water management. 2013;125:46-60.

[24] Neina D. The role of soil pH in plant nutrition and soil remediation. Applied and Environmental Soil Science. 2019.

[25] Mishra A, Tanna B. Halophytes: potential resources for salt stress tolerance genes and promoters. Frontiers in plant Science. 2017;8:829.

[26] Grigore MN, Ivanescu L, Toma C. Halophytes: An Integrative Anatomical Study. New York, Springer. 2014.

[27] Ola HAE, Reham EF, Eisa SS, Habib SA. Morpho-anatomical changes in salt stressed kallar grass (*Leptochloa fusca* L. Kunth). Research Journal of Agriculture and Biological Sciences. 2012;8:2:158-166.

[28] Bal AR, Dutt SK. Salt tolerance mechanism in *Acanthus Ilicifolius* L. Indian Journal of Plant Physiology. 1987;30:2:170-175.

[29] Flowers TJ, Flowers SA, Hajibagheri MA, Yeo AR. Salt tolerance in the halophytic wild rice, *Porteresia coarctata* Tateoka. New Phytologist. 1990;114:4:675-684.

[30] Cunxu W, Jianbo W, Yifang C, Weidong Z, Guorong S. Epicuticular wax of leaf epidermis: a functional structure for salt excretion in a halophyte {*Puccinellia tenuiflora*}. Acta Ecologica Sinica. 2004;24:11: 2451-2456.

[31] Abd Elhalim ME, Abo-Alatta OK, Habib SA, Abd Elbar OH. The anatomical features of the desert halophytes *Zygophyllum album* LF and *Nitraria retusa* (Forssk.) Asch. Annals of Agricultural Sciences. 2016;61:1: 97-104. DOI: 10.3390/agronomy7010018.

[32] Acosta-Motos JR, Ortuño MF, Bernal-Vicente A, Diaz-Vivancos P, Sanchez-Blanco MJ, Hernandez JA. Plant responses to salt stress: adaptive mechanisms. Agronomy. 2017;7:1:18.

[33] Hussain S, Khalid MF, Hussain M, Ali MA, Nawaz A, Zakir I, Fatima Z, Ahmad S. Role of micronutrients in salt stress tolerance to plants. In Plant nutrients and abiotic stress tolerance. 2018; Springer: Singapore: 363-376 p.

[34] Silva JV, Lacerda CFD, Costa PHAD, Enéas Filho J, Gomes Filho E, Prisco JT. Physiological responses of NaCl stressed cowpea plants grown in nutrient solution supplemented with CaCl2. Brazilian Journal of Plant Physiology. 2003;15:99-105.

[35] Huang J, Redmann RE. Physiological responses of canola and weld mustard to salinity and contrasting calcium supply. Journal of Plant Nutrition. 1995;18:9: 1931-1949.

[36] Razzaghi F, Ahmadi SH, Adolf VI, Jensen CR, Jacobsen SE, Andersen MN. Water relations and transpiration of quinoa (*Chenopodium quinoa* Willd.) under salinity and soil drying. Journal of Agronomy and Crop Science. 2011;197:5:348-360.

[37] Nicolas ME, Munns R, Samarakoon AB, Gifford RM. Elevated CO_2 improves the growth of wheat under salinity. Functional Plant Biology. 1993;20:3:349-360.

[38] Yadav T, Kumar A, Yadav RK, Yadav G, Kumar R, Kushwaha M. Salicylic acid and thiourea mitigate the salinity and drought stress on physiological traits governing yield in pearl millet-wheat. Saudi Journal of Biological Sciences. 2020;27:8:2010-2017.

[39] Agrawal RL. Fundamentals of plant breeding and hybrid seed production. Science Publishers, Inc. 1998.

[40] Silveira JAG, Araújo SAM, Lima JPMS, Viégas RA. Roots and leaves display contrasting osmotic adjustment mechanisms in response to NaCl-salinity in *Atriplex nummularia*. Environmental and Experimental Botany. 2009;66:1:1-8.

[41] Mahmood K. Salinity tolerance in barley (*Hordeum vulgare* L.): effects of varying NaCl, K?/Na? and $NaHCO_3$ levels on cultivars differing in tolerance. Pakistan Journal of Botany. 2011;43;3: 1651-1654.

[42] Qian YL, Engelke MC, Foster MJV. Salinity effects on zoysiagrass cultivars and experimental lines. Crop Science. 2000;40:2:488-492.

[43] Gulzar S, Khan MA, Ungar IA. Effects of salinity on growth, ionic content, and plant–water status of *Aeluropus lagopoides*. Communications in Soil Science and Plant Analysis. 2003;34:11-12:1657-1668.

[44] Jamil M, Rehman S, Rha ES. Response of growth, PSII photochemistry, and chlorophyll content to salt stress in four brassica species. Life Science Journal 2014;11:3:139-145.

[45] Zhen-hua ZHANG, Qiang LIU, Hai-xing SONG, Xiang-min RONG, Ismail AM. Responses of different rice (*Oryza sativa* L.) genotypes to salt stress and relation to carbohydrate metabolism and chlorophyll content. African Journal of Agricultural Research. 2012;7:1:19-27.

[46] Rai, V. Role of Amino Acids in Plant Responses to Stresses. Biologia Plantarum. 45, 481-487 (2002). DOI:10.1023/A:1022308229759

[47] Wang ZQ, Yuan YZ, Ou JQ, Lin QH, Zhang CF. Glutamine synthetase and glutamate dehydrogenase contribute differentially to proline accumulation in leaves of wheat (*Triticum aestivum*) seedlings exposed to different salinity. Journal of Plant Physiology. 2007;164:6:695-701.

[48] Silveira JA, de Almeida Viégas R, da Rocha IM, Moreira AC, de Azevedo

Moreira R, Oliveira JT. Proline accumulation and glutamine synthetase activity are increased by salt-induced proteolysis in cashew leaves. Journal of Plant Physiology. 2003;160:2:115-23.

[49] Hoshida H, Tanaka Y, Hibino T, Hayashi Y, Tanaka A, Takabe T, Takabe T. Enhanced tolerance to salt stress in transgenic rice that overexpresses chloroplast glutamine synthetase. Plant Molecular Biology. 2000;43:1:103-11.

[50] Santos C, Pereira A, Pereira S, Teixeira J. Regulation of glutamine synthetase expression in sunflower cells exposed to salt and osmotic stress. Scientia Horticulturae. 2004;103:1:101-11.

[51] Gao S, Ouyang C, Wang S, Xu Y, Tang L, Chen F. Effects of salt stress on growth, antioxidant enzyme and phenylalanine ammonia-lyase activities in *Jatropha curcas* L. seedlings. Plant, Soil and Environment. 2008;54:9:374-81.

[52] Valifard M, Mohsenzadeh S, Niazi A, Moghadam A. Phenylalanine ammonia lyase isolation and functional analysis of phenylpropanoid pathway under salinity stress in 'Salvia' species. Australian Journal of Crop Science. 2015;9:7:656-65.

[53] Jaarsma R, de Vries RS, de Boer AH. Effect of salt stress on growth, Na+ accumulation and proline metabolism in potato (*Solanum tuberosum*) cultivars. PloS One. 2013;8:3:e60183.

[54] Gholizadeh A, Kohnehrouz BB. Activation of phenylalanine ammonia lyase as a key component of the antioxidative system of salt-challenged maize leaves. Brazilian Journal of Plant Physiology. 2010;22:217-23.

[55] Sanada Y, Ueda H, Kuribayashi K, Andoh T, Hayashi F, Tamai N, Wada K. Novel light-dark change of proline levels in halophyte (*Mesembryanthemum crystallinum* L.) and glycophytes (*Hordeum vulgare* L. and *Triticum aestivum* L.) leaves and roots under salt stress. Plant and Cell Physiology. 1995;36:6:965-70.

[56] Arteaga S, Yabor L, Díez MJ, Prohens J, Boscaiu M, Vicente O. The use of proline in screening for tolerance to drought and salinity in common bean (*Phaseolus vulgaris* L.) genotypes. Agronomy. 2020;10:6:817.

[57] Khan N, Ali S, Zandi P, Mehmood A, Ullah S, Ikram M, Ismail MA, Babar M. Role of sugars, amino acids and organic acids in improving plant abiotic stress tolerance. Pakistan Journal of Botany. 2020;52:2:355-63.

[58] Morosan M, Al Hassan M, Naranjo MA, López-Gresa MP, Boscaiu M, Vicente O. Comparative analysis of drought responses in *Phaseolus vulgaris* (common bean) and *P. coccineus* (runner bean) cultivars. EuroBiotech Journal 2017;1:247-52.

[59] Borsai O, Hassan MA, Negrușier C, Raigón MD, Boscaiu M, Sestraș RE, Vicente O. Responses to salt stress in Portulaca: Insight into its tolerance mechanisms. Plants. 2020;9:12:1660.

[60] Poustini K, Siosemardeh A, Ranjbar M. Proline accumulation as a response to salt stress in 30 wheat (*Triticum aestivum* L.) cultivars differing in salt tolerance. Genetic Resources and Crop Evolution. 2007 Aug;54:5:925-34.

[61] De la Torre-González A, Montesinos-Pereira D, Blasco B, Ruiz JM. Influence of the proline metabolism and glycine betaine on tolerance to salt stress in tomato (*Solanum lycopersicum* L.) commercial genotypes. Journal of Plant Physiology. 2018;231:329-36.

[62] Agami RA. Applications of ascorbic acid or proline increase resistance to salt stress in barley seedlings. Biologia Plantarum. 2014;58:2:341-7.

[63] Kerepesi I, Galiba G. Osmotic and salt stress-induced alteration in soluble carbohydrate content in wheat seedlings. Crop Science. 2000;40:2:482-487.

[64] Ortega L, Fry SC, Taleisnik E. Why are *Chloris gayana* leaves shorter in salt-affected plants? Analyses in the elongation zone. Journal of Experimental Botany. 2006;57:14:3945-3952.

[65] Nandakumar MP, Shen J, Raman B, Marten MR. Solubilization of trichloroacetic acid (TCA) precipitated microbial proteins via NaOH for two-dimensional electrophoresis. Journal of Proteome Research. 2003;2:1:89-93. DOI: 10.1021/pr025541x. PMID: 12643547.

[66] Zhang M, Fang Y, Ji Y, Jiang Z, Wang L. Effects of salt stress on ion content, antioxidant enzymes and protein profile in different tissues of *Broussonetia papyrifera*. South African Journal of Botany. 2013;85:1-9.

[67] Mandhania S, Madan S, Sawhney V. Antioxidant defense mechanism under salt stress in wheat seedlings. Biologia Plantarum. 2006;50:2:227-231.

[68] Kaur S, Gupta AK, Kaur N. Gibberellin A3 reverses the effect of salt stress in chickpea (*Cicer arietinum* L.) seedlings by enhancing amylase activity and mobilization of starch in cotyledons. Plant Growth Regulation. 1998;26:2:85-90.

[69] Rohilla P, Yadav JP. Acute salt stress differentially modulates nitrate reductase expression in contrasting salt responsive rice cultivars. Protoplasma. 2019;256:5:1267-1278.

[70] Khan MG. Effect of salt stress on nitrate reductase activity in some leguminous crops. Indian Journal of Plant Physiology. 1994;37:3:185-187.

[71] Sharma PC, Gill KS. Effect of salinity on yield and ion distribution in pearl millet genotypes. Arid Land Research and Management. 1992;6:3: 253-260.

[72] Aly AAEH, Maraei RW, Aldrussi I. Changes in peroxidase and polyphenol oxidase activity and transcript levels of related genes in two Egyptian bread wheat cultivars (*Triticum aestivum* L.) affected by gamma irradiation and salinity stress. Bangladesh Journal of Botany. 2019;48:1:177-186.

[73] Özmen AD, Özdemír F, Türkan I. Effects of paclobutrazol on response of two barley cultivars to salt stress. Biologia Plantarum. 2003;4:2:263-268.

[74] Akhtar W, Mahmood T. Response of rice polyphenol oxidase promoter to drought and salt stress. Pakistan Journal of Botany. 2017;49:1:21-23.

[75] Plesa IM, González-Orenga S, Al Hassan M, Sestras AF, Vicente O, Prohens J, Sestras RE, Boscaiu M. Effects of drought and salinity on European Larch (*Larix decidua* Mill.) seedlings. Forests. 2018;9:6:320.

[76] Rasel M, Tahjib-Ul-Arif M, Hossain MA, Sayed MA, Hassan L. Discerning of rice landraces (*Oryza sativa* L.) for morpho-physiological, antioxidant enzyme activity, and molecular markers' responses to induced salt stress at the seedling stage. Journal of Plant Growth Regulation. 2020;39:1:41-59.

[77] Baniasadi F, Saffari VR, Moud AA. Physiological and growth responses of *Calendula officinalis* L. plants to the interaction effects of polyamines and salt stress. Scientia Horticulturae. 2018;234:312-7.

[78] Yang X, He K, Chi X, Chai G, Wang Y, Jia C, Zhang H, Zhou G, Hu R.

Miscanthus NAC transcription factor MlNAC12 positively mediates abiotic stress tolerance in transgenic *Arabidopsis*. Plant Science. 2018;277:229-41.

[79] Al-Huqail A, El-Dakak RM, Sanad MN, Badr RH, Ibrahim MM, Soliman D, Khan F. Effects of climate temperature and water stress on plant growth and accumulation of antioxidant compounds in sweet basil (*Ocimum basilicum* L.) leafy vegetable. Scientifica. 2020. DOI: 10.1155/2020/3808909

[80] Brenes M, Pérez J, González-Orenga S, Solana A, Boscaiu M, Prohens J, Plazas M, Fita A, Vicente O. Comparative studies on the physiological and biochemical responses to salt stress of eggplant (*Solanum melongena*) and its rootstock *S. torvum*. Agriculture. 2020;10:8:328.

[81] Caliskan O, Radusiene J, Temizel KE, Staunis Z, Cirak C, Kurt D, Odabas MS. The effects of salt and drought stress on phenolic accumulation in greenhouse-grown *Hypericum pruinatum*. Italian Journal of Agronomy. 2017;12:3.

[82] Rossi L, Borghi M, Francini A, Lin X, Xie DY, Sebastiani L. Salt stress induces differential regulation of the phenylpropanoid pathway in *Olea europaea* cultivars Frantoio (salt-tolerant) and Leccino (salt-sensitive). Journal of Plant Physiology. 2016;204:8-15.

[83] Attia H, Al-Yasi H, Alamer K, Esmat F, Hassan F, Elshazly S, Hessini K. Induced anti-oxidation efficiency and others by salt stress in *Rosa damascena* Miller. Scientia Horticulturae. 2020;274:109681.

[84] Cetinkaya H, Kulak M, Karaman M, Karaman HS, Kocer F. Flavonoid accumulation behavior in response to the abiotic stress: can a uniform mechanism be illustrated for all plants. Flavonoids-From Biosynthesis to Human Health. 2017;151-65.

[85] Shah A, Smith DL. Flavonoids in agriculture: Chemistry and roles in, biotic and abiotic stress responses, and microbial associations. Agronomy. 2020;10:8:1209.

[86] Di Ferdinando M, Brunetti C, Fini A, Tattini M. Flavonoids as antioxidants in plants under abiotic stresses. Abiotic stress responses in plants. 2012;159-79.

[87] Davies KM, Albert NW, Zhou Y, Schwinn KE. Functions of flavonoid and betalain pigments in abiotic stress tolerance in plants. Annual Plant Reviews Online. 2018;21-62.

[88] Pehlivan FE. Vitamin C: An antioxidant agent. Vitamin C. 2017;2:23-35.

[89] Muñoz P, Munné-Bosch S. Vitamin E in plants: biosynthesis, transport, and function. Trends in Plant Science. 2019;24:11:1040-51.

[90] Dietz KJ. Redox-dependent regulation, redox control and oxidative damage in plant cells subjected to abiotic stress. In Plant Stress Tolerance. Humana Press; 2010. 57-70 p.

[91] Kusvuran S, Kiran S, Ellialtioglu SS. Antioxidant enzyme activities and abiotic stress tolerance relationship in vegetable crops. Abiotic and biotic stress in plants-recent advances and future perspectives. 1st ed. Croatia: InTech; 2016. 481-503 p.

[92] You J, Chan Z. ROS regulation during abiotic stress responses in crop plants. Frontiers in Plant Science. 2015;6:1092.

[93] Hao Z, Wang X, Zong Y, Wen S, Cheng Y, Li H. Enzymatic activity and functional analysis under multiple abiotic stress conditions of a

dehydroascorbate reductase gene derived from *Liriodendron Chinense*. Environmental and Experimental Botany. 2019;167:103850.

[94] Jiang M, Zhang J. Water stress-induced abscisic acid accumulation triggers the increased generation of reactive oxygen species and up-regulates the activities of antioxidant enzymes in maize leaves. Journal of Experimental Botany. 2002;53:379:2401-10.

[95] Bray EA. Plant responses to water deficit. Trends in Plant Science. 1997;2:2:48-54.

[96] Wang W, Vinocur B, Altman A. Plant responses to drought, salinity and extreme temperatures: towards genetic engineering for stress tolerance. Planta. 2003;218:1:1-4.

[97] Metz J, Annighöfer P, Schall P, Zimmermann J, Kahl T, Schulze ED, Ammer C. Site-adapted admixed tree species reduce drought susceptibility of mature European beech. Global Change Biology. 2016;22:2:903-20.

[98] Sanjari S, Shobbar ZS, Ghanati F, Afshari-Behbahanizadeh S, Farajpour M, Jokar M, Khazaei A, Shahbazi M. Molecular, chemical, and physiological analyses of sorghum leaf wax under post-flowering drought stress. Plant Physiology and Biochemistry. 2021;159:383-91.

[99] Tuinstra MR, Grote EM, Goldsbrough PB, Ejeta G. Genetic analysis of post-flowering drought tolerance and components of grain development in *Sorghum bicolor* L. Moench. Molecular Breeding. 1997;3:6:439-48.

[100] Shavrukov Y, Kurishbayev A, Jatayev S, Shvidchenko V, Zotova L, Koekemoer F, de Groot S, Soole K, Langridge P. Early flowering as a drought escape mechanism in plants: How can it aid wheat production?. Frontiers in Plant Science. 2017;8:1950.

[101] Kooyers NJ. The evolution of drought escape and avoidance in natural herbaceous populations. Plant Science. 2015;234:155-62.

[102] Xu Z, Zhou G, Shimizu H. Plant responses to drought and rewatering. Plant signaling & behavior. 2010;5:6:649-54.

[103] Franks SJ. Plasticity and evolution in drought avoidance and escape in the annual plant *Brassica rapa*. New Phytologist. 2011;190:1:249-57.

[104] Aroca R, Porcel R, Ruiz-Lozano JM. Regulation of root water uptake under abiotic stress conditions. Journal of Experimental Botany. 2012;63:1:43-57.

[105] Willick IR, Lahlali R, Vijayan P, Muir D, Karunakaran C, Tanino KK. Wheat flag leaf epicuticular wax morphology and composition in response to moderate drought stress are revealed by SEM, FTIR-ATR and synchrotron X-ray spectroscopy. Physiologia Plantarum. 2018;162:3:316-32.

[106] Zhu X, Xiong L. Putative megaenzyme DWA1 plays essential roles in drought resistance by regulating stress-induced wax deposition in rice. Proceedings of the National Academy of Sciences. 2013;110(44):17790-5.

[107] Jaleel CA, Gopi R, Sankar B, Gomathinayagam M, Panneerselvam R. Differential responses in water use efficiency in two varieties of *Catharanthus roseus* under drought stress. Comptes Rendus Biologies. 2008;331:1 :42-7.

[108] Tsuji W, Ali ME, Inanaga S, Sugimoto Y. Growth and gas exchange of three sorghum cultivars under drought stress. Biologia Plantarum. 2003;46:4:583-7.

[109] Xiong J, Zhang L, Fu G, Yang Y, Zhu C, Tao L. Drought-induced proline

accumulation is uninvolved with increased nitric oxide, which alleviates drought stress by decreasing transpiration in rice. Journal of Plant Research. 2012;125:1:155-64.

[110] Gomez SM, Kumar SS, Jeyaprakash P, Suresh R, Biji KR, Boopathi NM, Price AH, Babu RC. Mapping QTLs linked to physio-morphological and plant production traits under drought stress in rice (*Oryza sativa* L.) in the target environment. American Journal of Biochemistry and Biotechnology. 2006;2:4:161-9.

[111] Khadka K, Earl HJ, Raizada MN, Navabi A. A physio-morphological trait-based approach for breeding drought tolerant wheat. Frontiers in Plant Science. 2020;11:715.

[112] Akman H, Zhang C, Ejeta G. Physio-morphological, biochemical, and anatomical traits of drought-tolerant and susceptible sorghum cultivars under pre-and post-anthesis drought. Physiologia Plantarum. 2021;172:2:912-21.

[113] Belay GA, Zhang Z, Xu P. Physio-Morphological and Biochemical Trait-Based Evaluation of Ethiopian and Chinese Wheat Germplasm for Drought Tolerance at the Seedling Stage. Sustainability. 2021;13:9:4605.

[114] Sdouga D, Amor FB, Ghribi S, Kabtni S, Tebini M, Branca F, Trifi-Farah N, Marghali S. An insight from tolerance to salinity stress in halophyte *Portulaca oleracea* L.: Physio-morphological, biochemical and molecular responses. Ecotoxicology and Environmental Safety. 2019;172:45-52.

[115] Borsai O, Hassan MA, Negrușier C, Raigón MD, Boscaiu M, Sestraș RE, Vicente O. Responses to salt stress in *Portulaca*: Insight into its tolerance mechanisms. Plants. 2020;9:12:1660.

[116] Vetter VM, Kreyling J, Dengler J, Apostolova I, Arfin-Khan MA, Berauer BJ, Berwaers S, De Boeck HJ, Nijs I, Schuchardt MA, Sopotlieva D. Invader presence disrupts the stabilizing effect of species richness in plant community recovery after drought. Global Change Biology. 2020;26:6:3539-51.

[117] Abid M, Tian Z, Ata-Ul-Karim ST, Wang F, Liu Y, Zahoor R, Jiang D, Dai T. Adaptation to and recovery from drought stress at vegetative stages in wheat (*Triticum aestivum*) cultivars. Functional Plant Biology. 2016;43:12:1159-69.

[118] Jangpromma N, Thammasirirak S, Jaisil P, Songsri P. Effects of drought and recovery from drought stress on above ground and root growth, and water use efficiency in sugarcane (*Saccharum officinarum* L.). Australian Journal of Crop Science. 2012;6:8:1298-304.

[119] Vesk PA, Westoby M. Drought damage and recovery: A conceptual model. New Phytologist. 2003;7-14.

[120] Saba J, Moghadam M, Ghasemi K, Nishabouri MR. Genetic properties of drought resistance indices. Journal of Agricultural Science and Technology. 2001;3:1:43-49.

[121] Nevo E, Chen G. Drought and salt tolerances in wild relatives for wheat and barley improvement. Plant, Cell & Environment. 2010;33:4:670-85.

[122] Oliver MJ, Cushman JC, Koster KL. Dehydration tolerance in plants. InPlant stress tolerance: Humana Press; 2010. 3-24 p.

[123] Barnabás B, Jäger K, Fehér A. The effect of drought and heat stress on reproductive processes in cereals. Plant, Cell & Environment. 2008;31:1:11-38.

[124] Xu W, Rosenow DT, Nguyen HT. Stay green trait in grain sorghum:

relationship between visual rating and leaf chlorophyll concentration. Plant Breeding. 2000;119:4:365-7.

[125] Thomas H, Ougham H. The stay-green trait. Journal of Experimental Botany. 2014;65:14:3889-900.

[126] Hilli HJ. Genetics and Genomic basis of Stay-green Trait. International Journal of Agriculture, Environment and Biotechnology. 2021;14:2:119-22.

[127] Al-Yasi H, Attia H, Alamer K, Hassan F, Ali E, Elshazly S, Siddique KH, Hessini K. Impact of drought on growth, photosynthesis, osmotic adjustment, and cell wall elasticity in Damask rose. Plant Physiology and Biochemistry. 2020;150:133-9.

[128] Khalil F, Naiyan X, Tayyab M, Pinghua C. Screening of EMS-induced drought-tolerant sugarcane mutants employing physiological, molecular and enzymatic approaches. Agronomy. 2018;8:10:226.

[129] Devi K, Prathima PT, Gomathi R, Manimekalai R, Lakshmi K, Selvi A. Gene expression profiling in sugarcane genotypes during drought stress and rehydration. Sugar Tech. 2019;21:5:717-33.

[130] ElBasyoni I, Saadalla M, Baenziger S, Bockelman H, Morsy S. Cell membrane stability and association mapping for drought and heat tolerance in a worldwide wheat collection. Sustainability. 2017;9:9:1606.

[131] Arteaga S, Yabor L, Díez MJ, Prohens J, Boscaiu M, Vicente O. The use of proline in screening for tolerance to drought and salinity in common bean (*Phaseolus vulgaris* L.) genotypes. Agronomy. 2020;10:6:817.

[132] Ahmed K, Shabbir G, Ahmed M, Shah KN. Phenotyping for drought resistance in bread wheat using physiological and biochemical traits. Science of the Total Environment. 2020;729:139082.

[133] Kumar A, Sharma A, Sharma R, Choudhary A, Srivastava P, Kaur H, Padhy AK. Morpho-physiological evaluation of *Elymus semicostatus* (Nees ex Steud.) Melderis as potential donor for drought tolerance in Wheat (*Triticum aestivum* L.). Genetic Resources and Crop Evolution. 2021;10:1-20.

[134] Haussmann BI, Hess DE, Omanya GO, Folkertsma RT, Reddy BV, Kayentao M, Welz HG, Geiger HH. Genomic regions influencing resistance to the parasitic weed *Striga hermonthica* in two recombinant inbred populations of sorghum. Theoretical and Applied Genetics. 2004;109:5:1005-16.

[135] Ejeta G, Tuinstra MR, Grote EM, Goldsbrough PB. Genetic analysis of pre-flowering and post-flowering drought tolerance in *Sorghum*. Production in Water-Limited Environments. 2000:137.

[136] Kebede H, Subudhi PK, Rosenow DT, Nguyen HT. Quantitative trait loci influencing drought tolerance in grain sorghum (*Sorghum bicolor* L. Moench). Theoretical and Applied Genetics. 2001;103:2-3:266-76.

[137] Alam MM, Mace ES, Van Oosterom EJ, Cruickshank A, Hunt CH, Hammer GL, Jordan DR. QTL analysis in multiple sorghum populations facilitates the dissection of the genetic and physiological control of tillering. Theoretical and Applied Genetics. 2014;127:10:2253-66.

[138] Bekele WA, Fiedler K, Shiringani A, Schnaubelt D, Windpassinger S, Uptmoor R, Friedt W, Snowdon RJ. Unravelling the genetic complexity of sorghum seedling development under low-temperature

conditions. Plant, Cell & Environment. 2014;37:3:707-23.

[139] Wang AY, Li Y, Zhang CQ. QTL mapping for stay-green in maize (*Zea mays*). Canadian Journal of Plant Science. 2012;92:2:249-56.

[140] Sreeman SM, Vijayaraghavareddy P, Sreevathsa R, Rajendrareddy S, Arakesh S, Bharti P, Dharmappa P, Soolanayakanahally R. Introgression of physiological traits for a comprehensive improvement of drought adaptation in crop plants. Frontiers in Chemistry. 2018;6:92.

[141] Amelework B, Shimelis H, Tongoona P, Laing M. Physiological mechanisms of drought tolerance in sorghum, genetic basis and breeding methods: a review. African Journal of Agricultural Research. 2015;10:31: 3029-40.

Chapter 3

Salt Stress Responses and Tolerance in Soybean

Mirza Hasanuzzaman, Khursheda Parvin, Taufika Islam Anee, Abdul Awal Chowdhury Masud and Farzana Nowroz

Abstract

Soybean is one of the major oil crops with multiple uses which is gaining popularity worldwide. Apart from the edible oil, this crop provides various food materials for humans as well as feeds and fodder for animals. Although soybean is suitable for a wide range of soils and climates, it is sensitive to different abiotic stress such as salinity, drought, metal/metalloid toxicity, and extreme temperatures. Among them, soil salinity is one of the major threats to soybean production and the higher yield of soybean is often limited by salt stress. Salt stress negatively affects soybean seedling establishment, growth, physiology, metabolism, and the ultimate yield and quality of crops. At cellular level, salt stress results in the excess generation of reactive oxygen species and creates oxidative stress. However, these responses are greatly varied among the genotypes. Therefore, finding the precise plant responses and appropriate adaptive features is very important to develop salt tolerant soybean varieties. In this connection, researchers have reported many physiological, molecular, and agronomic approaches in enhancing salt tolerance in soybean. However, these endeavors are still in the primary stage and need to be fine-tuned. In this chapter, we summarized the recent reports on the soybean responses to salt stress and the different mechanisms to confer stress tolerance.

Keywords: abiotic stress, antioxidant defense, climate change, *Glycine max*, NaCl, osmotic stress, oxidative stress, reactive oxygen species, sustainable agriculture

1. Introduction

Soybean (*Glycine max* L.) is considered to be one of the major oilseed crops as well as an economically important leguminous crop as it supplies more than 25% of the global protein requirement for food and feed [1, 2]. Not only the food and feed values but also the root nodules of soybean enhance soil fertility through nitrogen (N) fixation. It also has other uses like forage crops, emergency crops, and even as a source of bioenergy [3]. A crop with such superior qualities needs to gain more attention than it has currently. Soybean production has been increasing over years, but the rate is not as much as other major cultivated crops like maize, wheat, and rice. The possible reasons are lack of suitable varieties, different environmental constraints, etc.

Environmental stresses including salinity, drought, waterlogging, toxic metals, extreme temperatures, etc. are nowadays the center of concern by the environmentalist and plant scientists worldwide as these are badly affecting food production. Salt stress is undoubtedly one of the worst conditions for plant growth as it creates both osmotic and ionic stress. Soybean is known to be partially sensitive to salt which may result in up to 40% yield loss depending upon salinity level. The presence of excess salt in the growing medium of soybean negatively affects the quality and quantity of seed, growth, and nodulation process [4]. Synthesis of protein, uptake and transportation of water and nutrient, translocation of assimilates, cytosolic and mitochondrial reactions, and several other metabolic pathways are adversely affected by salt stress [5]. Dehydration of cell and toxic ion accumulation occur when the rhizobia-legume symbiosis process is hampered in particular [4]. Higher magnitude of salt stress may even cause plant death [6]. In addition to these, salt stress impairs the photosynthesis process that ultimately results in oxidative stress by the excess production of reactive oxygen species (ROS). Ionic imbalances disrupt the normal metabolic processes and subsequently aggravate ROS generation making the situation worse for plants [5, 7, 8]. The antioxidant defense system is a vital strategy for the alleviation of oxidative stress in plants. It consists of different enzymatic and non-enzymatic components which act as saviors at their optimum levels. Besides, the use of salt-tolerant varieties, seed priming, use of exogenous protectants, beneficial microbes, or proper modulation of agronomic managements are other potential strategies to enhance salt stress tolerance in different crops including soybean.

In the present situation of the world where hunger and malnutrition exist, people are badly in need of proper nutrition to ensure a healthy future. In that case, soybean being a crop of versatile nutritional values and wide adaptability should be focused on the agriculture sector worldwide. But, due to some existing challenges the per unit yield of soybean is not increasing, although its global production has been increased remarkably over years. The underlying reason is the lack of proper knowledge about the identification of yield traits that could facilitate the per unit yield of soybean. Unlike cereal crops, soybean architecture is more complex, which makes the task by the breeders difficult to optimize the plant accessions. For example, simply increasing plant height will not give a higher yield due to lodging, and dwarf varieties will also reduce yield due to fewer internodes. Hence, Liu et al. [2] suggested that lodging resistant long plants with shorter internodes might be helpful in this regard. They also suggested that minimization of pod abscission is also required to ensure a better yield. So, it is clarified that high-yielding soybean variety development is not a piece of cake. But it can be made possible by an in-depth understanding of soybean plant architecture and yield characteristics. Moreover, environmental adversities like salinity are making these tasks more challenging as it hampers soybean growth and productivity by inhibiting germination and other vital physiological processes [3]. However, soybean is known to be a moderately salt-sensitive crop, and therefore, to ensure the potential production of soybean we need to minimize the salt stress-induced damages in soybean. To execute that, extensive studies on soybean responses to the different magnitude of salt stress and possible adaptive mechanisms are necessary. In this chapter, we have summarized the up-to-date findings focused on the responses of soybean to salt stress and exploring the tolerance mechanisms.

2. Soybean: a crop of multiple uses

Soybean is a crop of versatile uses with wide adaptability. Apart from the soybean oil, the whole soybean including pod and seed have multifarious uses

Figure 1.
Versatile uses of soybean products [adapted from Hasanuzzaman et al. [3] with permission from Elsevier].

like human food, animal feed, and in improving soil fertility through N-fixation (**Figure 1**). Soybean oil and protein contents are 21 and 40%, respectively including carbohydrate and ash contents of at 34 and 5%, respectively [9]. As a vegetable, early maturing soybean variety with green and immature pods is more feasible for fresh or frozen consumption [10]. Food products made from soybean e.g., soya milk, soya flour, tofu, boiled soybean, soya meat, etc. are gaining popularity day by day for their nutritional values. Soybean germ oil, produced as a byproduct at the time of protein preparation is able to reduce plasma cholesterol and so can be used in the treatment of hypercholesterolemia [11]. Having high concentrations of β-carotene, vitamin C, high calories and essential amino acid soybean also can be efficiently used in anti-hypertensive, antimicrobial, antioxidant, anti-diabetic, and anticancer activities that are very important for human health [12, 13]. Moreover, soybean is now being used in the preparation of meat analog, which has similarities with meat but is more beneficial in lowering lipid and blood pressure while increasing low-density lipoprotein (LDL), cholesterol oxidation, and digestibility [14].

A rich amount of protein content with optimum amino acid profile, low amount of crude fiber, high phosphorus (P) content, and high level of digestibility is noticeable at soybean meal that makes it a desirable feed for livestock [15]. Whole plants, raw seeds, or processed products of soybean can be used efficiently in cattle feeds if proper ration components can be chosen considering the rate of N and energy release during rumen fermentation processes [16]. Additionally, soybean oil, due to its rich fatty acid profile, can be a convenient and affordable aquafeed in the aquaculture industry around the world [17, 18].

Soybean is a well-fitted crop in a wide range of climates and soils. It is easily cultivated as an intercrop, facilitates in proper utilization of available resources, and soil fertility improvement through biological N-fixation. When crop rotation was done comprising corn and soybean, minimum nutrient and water loss with higher yield and net income was recorded compared to conventional farming [19, 20].

Nowadays, this soybean solid waste is being widely used for different purposes, especially in bioenergy. If soybean waste is managed through proper hydrothermal treatment, can be efficiently used in industries as an eco-friendly way for its good nutritional, thermal, and fuel properties [21]. Soybean meal when used with bark, can be a raw material to prepare an affordable, formaldehyde-free, and environmentally sound wood adhesive for plywood production, which has a better performance in giving high water resistance, mechanical strength, and thermal stability [22, 23]. However, soybean could be a potential renewable energy source for different industrial purposes.

3. Soybean response to salt stress

3.1 Seed germination and emergence

Uniform seed germination and rapid seedling growth are prerequisites for better crop establishment [24, 25]. Salinity hampers the germination process either by delaying or reducing the germination rate [26]. Simaei et al. [27] observed that at lower concentrations of salt stress (0.05 and 0.1%) soybean seed germination was delayed whereas, at high salt dose germination rate was reduced. Shu et al. [28] reported that germination of soybean seed was delayed under 150 mM NaCl stress. Additionally, post germinative growth of soybean seedlings was adversely affected due to salinity. Salt stress downregulates gibberellic acid (GA) and upregulates abscisic acid (ABA) bio-production that ultimately reduces the GA/ABA ratio and thus delays seed germination in soybean [28]. Soybean seedling growth completely stopped when the leaf sodium (Na^+) reached 6.1 mg g^{-1} fresh weight (FW). It was also observed that compared to the control, seedling growth reduced by 14% at 160 mM NaCl whereas, at 330 mM NaCl stress soybean seedling growth got completely stopped [29]. Delayed seed germination processes in three cultivars of soybean in terms of germination rate, radicle length, and FW of germinated seeds were observed by Shu et al. [28]. During the seed germination process, changes in cell oxidative status were reported under salt stress. Xu et al. [30] conducted an experiment with two cultivars of soybean viz., Lee68, and N2899 to evaluate the response of soybean seeds to salt stress during germination. Seeds were soaked in 100 mM NaCl until radical initiation and reported that without impacting final germination, salt only delayed mean germination time (MGT) by 0.3 and 1.0 d in Lee68 and N2899, respectively. Abulfaraj and Jalal [31] experimented with three cultivars (Crawford, G-111, and Clark) of soybean viz., under 200 and 400 mM NaCl stress and observed that the final germination percentage (FGP) was reduced in all three cultivars irrespective of stress levels. However, it was prominent that the higher the stress, the lower was the FGP. At 400 mM salt stress the highest reduction of FGP (60%) was observed in cultivar Clark, whereas Crawford and G-111 showed 29 and 22% reduction, respectively. However, the MGT was increased with increasing salinity level where Crawford exhibited the best performance by lowering MGT values than other cultivars under both salinity stress and normal condition [31]. Essa [32] observed a remarkable reduction of germination percentage under different doses of salt stress in three soybean cultivars. They found that, when salinity level increased from 0.5 to 8.5 dS m^{-1}, it reduced germination by 54 and 63% for Lee and Coiquitt cultivars, respectively. Hashem et al. [33] observed that in two soybean cultivars, seed germination was also reduced with increasing salt doses. At 200 and 300 mM NaCl, germination reduced by 4 and 11%, respectively for the salt-tolerant variety Clark, whereas 11 and 20% were documented in salt-sensitive variety Kint, respectively. Khan et al. [34] experimented with different doses of NaCl treatments ranges (50–300 mM) and revealed that above 200 mM salt concentration, the maximum seed did not germinate whereas, most of the seeds germinated under the 50 and 100 mM NaCl treatments.

3.2 Growth

One of the obvious effects of salt stress in soybean is the reduction of plant growth, which is reported in many plant studies [28, 35, 36]. As salt stress inhibits water and nutrient uptake and translocation, it is obvious that this hampers the normal cell growth and development and hence, growth is retarded. Otie et al. [37] found that plant height, number of leaves, leaf area index (LAI) and the specific leaf area (SLA) of soybean plants were declined in a dose-dependent manner when they were grown under salinity (32.40, 60.60 and 86.30 mM NaCl). Salt stress also resulted in a lower dry weight (DW) of the leaf and stem [37]. The higher salinity declined the cell elongation and division that may lead to the inhibition of growth-related metabolic and physiological processes and eventually restricted the allocation of biomass [38].

Salt stress also hampers the plant's anatomy and impairs the normal growth of plants. As salt stress allows the entry of high Na^+ into plants roots, this reduces the elongation rate and disturbs root architecture [35, 36, 39]. It is evident that plants sometimes thicken the epidermis and endodermis root cells as a preventive measure of Na^+ influx. But under severe salt stress, this phenomenon may lead to cell expansion and cell wall integrity [40, 41]. Silva et al. [36] observed that Na^+ stress differently affected root anatomy. Interestingly, lower (50 and 100 mM Na^+) salt stress improved the root epidermis and endodermis thickness, cortex thickness, vascular cylinder diameter, and metaxylem diameter at different root depths, while higher (150 and 200 mM Na^+) salt stress reduced these parameters. Root epidermis and endodermis thickness were decreased by 44 and 56% under 200 mM Na^+, compared to the control treatment, while root cortex thickness and vascular cylinder diameter were decreased by 8–17%, respectively [36]. Salt-induced decline of plant growth exclusively depends on the salinity levels and the duration of exposure. At the initial stage, salt-induced decline in seedling growth is associated with the seedling vigor and seed germination rate which is due to lower synthesis of particular phytohormones such as GA.

Amirjani [42] recorded some reduction of plant height and FW of soybean plants under salinity stress. A reduction of plant height of 30, 47, and 76% and FW of 32, 54, and 76% were found by increasing salinity levels to 50, 100, and 200 mM, respectively. Weisany et al. [43] observed that salinity stress caused a number of morphological and physiological changes in soybean plants. Salt stress caused a decrease in shoot length, root FW and DW, and shoot FW and DW. Lee et al. [44] found a marked decline (27%) in shoot length when soybean plants were exposed to 80 mM NaCl. The soybean FW and DW were also significantly decreased with NaCl application which was 16 and 40%, respectively lower than control. In our experiment, we found that soybean plants exposed to 300 and 450 mM NaCl stress showed 35 and 55%, respectively reduction in plant height. Moreover, shoot FW and DW were declined by 43 and 41%, respectively at 150 mM NaCl stress [45].

3.3 Photosynthesis

It is well-established that yield reduction under salinity stress is due to the reduced production of photo-assimilate, slower transportation of photosynthetic components, and transformation in cytosolic metabolism [46]. Under salinity stress, the concentration of Na^+ ion absorption gets increased in plant tissues. As a result, uptake of other essential nutrients required for different biosynthesis processes gets decreased [47, 48]. Excessive salt accumulation in the cell causes necrosis and reduces the photosynthesis rate that ultimately diminishes plant growth [49]. In soybean, excess uptake of Na^+ under saline conditions decreases potassium (K^+)

uptake which is a vital regulator of stomatal opening and closing during photosynthesis. Furthermore, prolonged salt stress results in chlorosis and reduced content of different photosynthetic pigments in soybean [27, 50]. Various experiments conducted with different cultivars of soybean reported that plant photosynthetic parameters significantly varied with the duration and doses of salt stress (**Table 1**). The experiment conducted by Zaki et al. [58] found that total chlorophyll (chl), carotenoid contents, and chl fluorescence were decreased in 150 mM NaCl stress compared to control. Hashem et al. [33] experimented with two soybean cultivars (Clark and Kint) under 200 mM NaCl stress and observed considerable reductions in chl *a*, chl *b*, total chl, and carotenoids by 33, 35, 35, and 50% for Clark and 71, 76, 74, and 81% for Kint, respectively. Here it is evident that in tolerant cultivar the rate of photosynthetic reduction was lower compared to sensitive one. Significant reduction in all photosynthetic traits such as net photosynthetic rate, stomatal conductance, and internal CO_2 concentration was observed in soybean plants when exposed to 100 mM NaCl stress. Considering the duration, under 6 h of salt stress, net photosynthesis was reduced by 47% whereas, 81% reduction was observed at 48 h stress. Furthermore, at 24 h and 48 h stress durations, intercellular CO_2

Name of cultivars	Dose and duration of salt stress	Stress responses	Reference
Clark and Kint	200 mM NaCl, 60 d	Reduction of chl *a*, *b*, total chl and carotenoids contents by 33, 35, 35, and 50% for Clark and 71, 76, 74, and 81% for Kint cultivar, respectively	[33]
Giza 111	75 and 150 mM NaCl, 14 d	Reduction of chl *a*, *b*, *a* + *b*, and carotenoids by 13, 22, 16, and 7%, respectively at 75 mM salt stress and by 27, 23, 26, and 27%, respectively at 150 mM salt stress	[51]
C08	150 mM NaCl, 2d	Reduction of photosynthetic rate and stomatal conductance	[52]
Giza 22	100 mM NaCl, 25 d	Reduction of chl and carotenoid contents by 46 and 40%, respectively. Reduction of photosynthetic rate, intercellular CO_2 concentration, and stomatal conductance by 42, 26, and 22%, respectively	[53]
Giza 111	3 and 6 g L^{-1}, saline water (dissolving sea salt with tapwater), 30 d	Dose-dependent reduction of chl *a*, chl *b*, and carotenoid contents	[54]
M7	4, 7, and 10 dS m^{-1}, 60 d	Reduction of leaf chl, carotenoids, and anthocyanins contents Reduction of Fv/Fm at above 4 dS m^{-1} salinity	[55]
Crawford, Giza 111, Clark	200 and 400 mM NaCl, 69 d	Reduction of chl *a* content by 69, 79, and 85% in Crawford, Giza 111, and Clark, respectively, and chl *b* content by 61 and 71% in Crawford and Clark, respectively	[31]
Giza 22	7.46 dS m^{-1} saline soil, 60 d	Reduction of total chl and carotenoid contents	[56]
M7	5 and 10 dS m^{-1}, 30 d	Reduction of chl *a*, chl *b*, chl *a*/*b*, and total chl contents Reduction of RuBisCO activity with increasing salt concentration	[57]

Table 1.
Alteration of photosynthesis and associated parameters of soybean under salt stress.

concentration was decreased by 2 and 11%, respectively at the same dose of salinity stress [59]. In a similar experiment, Soliman et al. [53] observed that after 25 d of salt (100 mM NaCl) condition, stomatal conductance, chl and carotenoids contents, photosynthetic rate, and intercellular CO_2 concentration were reduced by 22, 46, 40, 42, and 26%, respectively. Yan et al. [60] experimented with a halophytic soybean (*Glycine soja*) to explore the salt adaptability in terms of photosystem coordination under 300 mM NaCl for 9 days (d). The result showed that under stress conditions, photosystem II (PSII) electron transport rate, stomatal conductance, and photosynthetic rate were reduced in both *G. soja* and *G. max*; however, the highest reduction was reported in *G. max*. Oppositely, PSII excitation pressure was increased by 72 and 50% in *G. max* and *G. soja,* respectively on day 9. During photosynthesis, chl plays a vital role in photon harvest in the PSII and PSI that belongs to the chloroplast. Under stress conditions, the production of different ROS is subjected to the destruction of many cellular organelles along with chloroplast. As a result, other photosynthetic activities are significantly hampered under salt stress [61]. Ning et al. [62] observed that under the same level of salt stress, sensitive cultivars showed higher photosynthesis inhibition compared to tolerant ones. It is assumed that due to the higher accumulation of P, K^+, and magnesium (Mg^{2+}) in the leaf, tolerant cultivar restored higher photosynthetic rates. When the Giza 111 cultivar of soybean was exposed to different levels of salt stress for 14 d, it resulted in a significant reduction of the photosynthetic pigments. Compared to control chl *a*, chl *b*, chl (*a* + *b*), and carotenoids contents were decreased by 27, 23, 26, and 27%, respectively [51]. Leaf size and area are two vital factors that regulate the amount of light captured for photosynthesis. Under salt stress, leaf area is reduced significantly and directly affects the production of photosynthetic pigments by reducing the amount of calcium (Ca^{2+}) and iron (Fe) ions in leaves responsible for the chl biosynthesis [63, 64]. All these results reported in soybean under salinity stress are the consequences of osmotic stress, which triggers the higher accumulation of Na^+ in leaf restricting the supply of K^+ and thus inhibits photosynthesis, which eventually reduced the stomatal conductance. In addition, this excessive ion toxicity reduces chl content, damages chloroplast structure, and leads to nonstomatal inhibition of photosynthesis [62]. Guo et al. [65] also suggested that high Na^+ notably reduces the K^+, Ca^{2+}, Mg^{2+}, and other cations in the leaf which are vital for the photosynthesis in the plant.

3.4 Water relation

One of the major effects of soil salinity is the osmotic stress in plants. Under salt stress, the water uptake is hampered due to the lack of energy and also the imbalance between solute concentration in the soil solution and plant cells. As a result, plant cells lose their turgidity which results in osmotic stress. Katerji et al. [66] reported that water consumption declined gradually as the salinity increased. As a result, the water content of the leaves was reduced and with it the turgor. Stomatal closure and eventually transpiration and photosynthesis were reduced resulting in growth retardation. The leaf relative water content (RWC), xylem exudation, leaf water potential were evidently declined in soybean plants subjected to 75 mM NaCl [34]. This decline in RWC indicates a loss of turgor which is associated with impaired water availability required for cell growth and development [67]. Shoot water content and leaf water potential were decreased by salinity [68]. Ferdous et al. [69] conducted an experiment with soybean under salt stress and found that RWC was decreasing with increasing salinity levels. The control treatment showed 89% RWC and it reduced to 73% at 100 mM salinity level. Two soybean genotypes viz. Shohag and AGS 313 were tested against salinity (50 and 100 mM NaCl) for different durations (15, 30, 45, 60, and 80 d) and it was observed that salt stress caused a reduction in RWC, water retention capacity, leaf water potential, and exudation rate in a concentration and duration-dependent manner [70]. The decline in exudation rate indicated the lower flow of water into plants which is associated with lower water potential and eventually RWC. However, the moderately tolerant genotype i.e., AGS 313 showed relatively higher RWC, water retention capacity, leaf water potential, and ER compared to the cultivar Shohag. One of the obvious effects of salt stress in soybean is the reduction of osmotic potential. A dose-dependent decline of osmotic potential was observed in soybean plants upon 28-d exposure to salt (60 and 120 mM NaCl), with more negative values in the treatment with 120 mM NaCl [71].

3.5 Yield and quality

Exposure of plants to salt conditions causes morphological, physiological, and biochemical alterations in the plant, which ultimately pose negative impacts on plant reproductive attributes and subsequent yield [72]. The primary effect of salt stress is a shortage of water, which is a crucial element for soybean flowering and podding. Upon exposure to salt stress, soybean growth attributes (shoot length, stem diameter, number of branches, flower number, pod number, and seed weight) were recorded to decrease with a higher accumulation of Na^+ in leaves (**Table 2**). These changes could be the reason behind the manifestation of reduced yield and oil content of soybean [55, 84]. Additionally, the osmotic effect of salt stress influences the augmentation of growth retardation, obliteration of growth

Name of cultivars	Stress level and duration	Effects on yield and quality	References
Peking	250 mM NaCl	The number of flowers and pods was decreased by 64 and 69%, respectively Pod length and weight were reduced by 10 and 20%, respectively. Reduction of seed yield by 55%	[73]
Giza 111	104.44 mM Sea salt, 30 d	Pod number and weight were decreased by 31 and 60%, respectively Seeds number and weight were reduced by 41 and 72%, respectively	[54]
Williams	9 dS m^{-1}, 75 d	Reduction of grain, protein, and oil yield $plant^{-1}$ by 39, 38, and 39%, respectively	[74]
JS-335	100 mM NaCl, 45 d	The number of pods $plant^{-1}$ and harvest index were greatly reduced by about 51 and 18%, respectively	[75]
JS-335	50 mM NaCl, 45 d	The number of pods $plant^{-1}$, seeds, and seed weight were decreased by about 29, 18, and 17%, respectively	[76]
M7	10 dS m^{-1}, 70 d	Pods and seeds $plant^{-1}$ were reduced by 28 and 30%, respectively	[55]
M7	10 dS m^{-1}, 109 d	Seed yield reduction by 44% Soluble sugar and protein contents were increased by 389 and 108%, respectively	[77]

Name of cultivars	Stress level and duration	Effects on yield and quality	References
Sohag	50 mM NaCl	Pods plant^{-1} and seeds pod^{-1} were reduced by 52 and 10%, respectively Reduction of seeds and yield plant^{-1} by 54 and 51%, respectively 100-seed weight was decreased by 16%	[46]
Hwangkeumkong	140 mM NaCl, 40 d	Pods plant^{-1} and pod DW were reduced by 40 and 78%, respectively 100-seed weight and yield were decreased by 55 and 80%, respectively	[78]
Giza 22	7.46 dS m^{-1}, 70 d	The number of pods plant^{-1}, 100-seed weight, and seed yield were reduced by about 59, 57, and 55%, respectively Reduction of protein and oil contents by 30 and 41%, respectively	[56]
Giza 111	150 mM NaCl, 39 d	Pod number plant^{-1} and 100-seed weight were decreased by 40 and 45%, respectively Seed protein and oil contents were decreased by 24 and 29%, respectively Increased soluble sugar content by 17%	[58]
Pungsannamul	300 mM NaCl, 21 d	Increment of total protein content by about 57%	[79]
Giza 35	80 mM NaCl, 5 weeks	Soluble protein and sugar content were reduced by 26 and 24%, respectively	[80]
Giza 22	9.4 dS m^{-1}, 8 weeks	Pod number plant^{-1} and 100-seed weight were decreased by 28 and 34%, respectively Seed weight plant^{-1} and yield were reduced by 27 and 27%, respectively Reduction of protein and oil content at 19 and 11%, respectively	[81]
No. 62	12 dS m^{-1}	Pods plant^{-1} was reduced by 54% Protein and oil content were decreased by 24 and 10%, respectively	[82]
Galarsum, BD 2331, and BARI Soybean 6	50 and 75 mM NaCl, 35 d	The number of pods plant^{-1} was reduced with the lowest in BD2331 (38% compared to control) at 75 mM NaCl The number of filled pods plant^{-1} was lowest in BARI Soybean 6 (18% compared to control)	[83]
Tachiyutaka8428h	86.3 mM, 2 weeks	Days to 50% podding required 39% more time than control plants The number of pods plant^{-1} reduced by 16.5% compared to the control plants	[37]

Table 2.
Changes in yield and quality attributes of soybean under salt stress.

promoter, imbalanced ions, water uptake, and inhibition of photosynthetic activities that eventually alter the growth traits responsible for yield, yield attributes, oil and protein content of seed [56, 85]. Soybean productivity depends on the root performance under different soil conditions. As salt-induced condition disturbs the nodule formation, impeding the activity of root, yield reduction is quite obvious here [23]. Salt stress minimizes the grain size along with the duration of protein and oil accumulation, in turn, decreasing oil and protein content [74, 82]. Decreased protein content also can be occurred due to the interruption of N metabolism and nitrate absorption within the plant, which is a common phenomenon under salt stress [86]. Under severe salt stress (80 mM NaCl) both the soluble sugar and soluble protein that are helpful in maintaining osmotic adjustment are recorded to decrease [80]. But the changes in sugar content under salt-induced conditions may vary according to soybean varieties. Tolerant variety may show increased sugar concentrations under salt stress to maintain turgor within plant species [87].

However, yield attributes e.g., branch number, pod number, number of seeds, etc. performed variably in different genotypes [46]. In addition, the duration of salt exposure also can influence the level of productivity loss. Pre-flowering stage of soybean is recorded to be more susceptive to salt stress as it reduces the capacity of soybean to uptake water and nutrient by accumulating a higher concentration of Na^+ in the root [78].

3.6 Oxidative stress in soybean under salinity

Salt stress-induced stomatal closure leads to the reduction of the availability of CO_2 in leaves which reduces carbon fixation. Hence, the chloroplasts get exposed to higher excitation energy resulting in overproduction of ROS that includes superoxide ($O_2^{\bullet-}$), hydrogen peroxide (H_2O_2), hydroxyl radical ($OH^{\bullet}$), and singlet oxygen ($^1O_2$), etc. Besides, either osmotic or ionic stress resulting from salt exposure ultimately provokes the levels of ROS leading to cellular damages through the oxidation of lipids, proteins, and nucleic acids. The plant itself possesses some ROS scavenging mechanisms, such as an antioxidant defense system, and when there is an imbalance between the level of ROS production and the antioxidant defense system, the plant fails to cope with the oxidative stress (**Figure 2**) [72].

Recent reports demonstrated the salt stress-induced oxidative damages in soybean plants (**Table 3**) and a notable number of those have also included possible mitigation strategies. For example, soybean seedlings grown on a pot containing

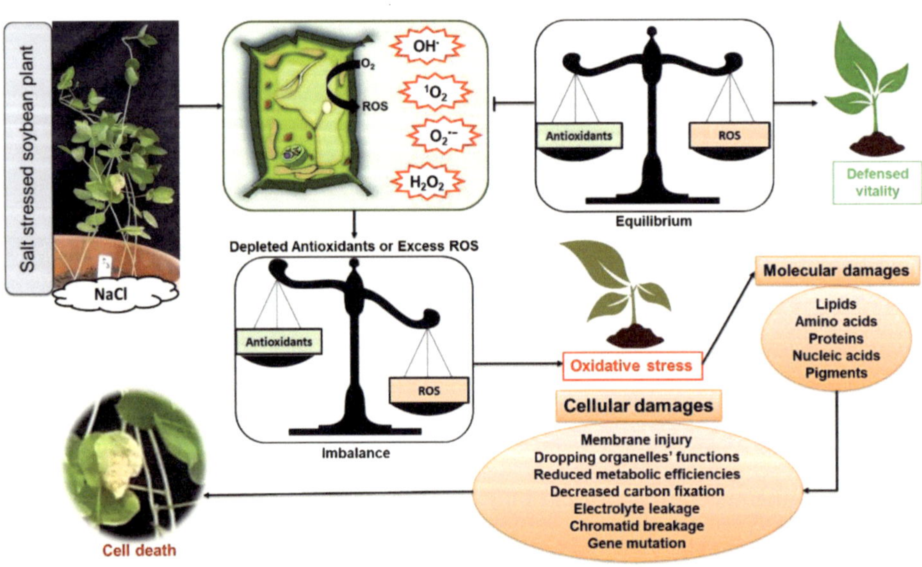

Figure 2.
Generalized outline of salt-induced oxidative stress and its consequences in soybean plants [adapted from Hasanuzzaman et al. [88].

Name of cultivars	Salinity levels	Level of oxidative stress	Reference
Taegwang	100 mM NaCl, 2 d	Levels of both lipid peroxidation and H_2O_2 content were higher	[89]
Giza 35	80 mM NaCl, 5 weeks	2.7-fold increase of MDA and 3.7-fold increase of H_2O_2 contents were recorded	[80]
Giza 111	150 mM NaCl, 40 d	EL was two times higher and MSI was 25% lower	[58]
Giza 22	7.46 dS m^{-1}, 70 d	A notable increase in MDA and H_2O_2 contents was recorded	[56]
FD92 and Z1303	150 mM NaCl, 20 d	MDA and H_2O_2 contents were 62 and 122% higher, respectively	[90]

Table 3.
Oxidative stress responses of soybean to salt stress.

100 mM NaCl accumulated 38% higher H_2O_2 and caused 25% higher lipid peroxidation compared to the seedlings grown with no NaCl [91]. Exposure to salinity (6 dS m^{-1}) can also raise electrolyte leakage (EL) by 69% along with 75% and 56% increase in H_2O_2 and malondialdehyde (MDA) contents, respectively in soybean plants [92]. A similar increment of these oxidative stress markers in a dose-dependent manner of NaCl (75 and 100 mM) was also reported by Alharby et al. [5]. Reduction in membrane stability index (MSI) denotes the loss of cell membrane integrity and 21% lower MSI in soybean plants was recorded upon exposure to 100 mM NaCl stress [93]. They also observed 2.74 and 3.94-fold higher ROS ($O_2^{\bullet-}$ and H_2O_2, respectively) production and 2.28-fold higher lipid peroxidation in soybean. In another experiment with two cultivars of soybean differing in their salt tolerance levels (tolerant, Nannong 99-6 and sensitive, Lee 68), it was depicted that the oxidative or cellular damages are usually higher in the salt-sensitive one upon exposure to salinity [94]. Klein et al. [95] assessed the contents of H_2O_2 and MDA in root nodules of soybean exposed to 70 mM NaCl for 12 d and noted almost 98 and 75% enhancement of H_2O_2 and MDA, respectively.

Being one of the catastrophic abiotic stresses, salt stress induces oxidative stress in many possible ways including disrupted stomatal conductance, intruded photosynthesis, and altered activities of different enzymes [96]. The above-mentioned increase of lipid peroxidation, ROS production, EL, or reduction of MSI accounts for salt-induced oxidative damages in soybean plants. The bright side is that there are a number of mitigation strategies or adaptive mechanisms available for plants that facilitate the recovery or protection of plant cells from the cellular damages caused by excess salt.

4. Enhancing salt tolerance and adaptative mechanisms in soybean

4.1 Using salt tolerant genotypes

With the advancement of modern science and inventions in breeding technology; scientists, breeders, and agriculturists are working relentlessly to develop more tolerant varieties in different crops through genomic approaches and inbreeding technologies. Approaches towards genetic modification of soybean genotypes in order to increase their performances under salt-affected soil are of great importance. To identify salt-tolerant genotypes through a successful breeding program, the genetic diversity present in different soybean germplasm needs to be evaluated through a proper screening process that will help in understanding the mechanisms of salt tolerance in diverse genotypes [97]. United States Department of Agriculture (USDA) has compiled the salt tolerance level of more than 550 genotypes till today. Among these, 151 were declared as tolerant and 413 as susceptible genotypes [98]. The most effective way to uphold soybean yield under salinity stress is genetic improvement of the existing soybean varieties. To do so, in the beginning, requires the identification of genetic traits in soybean germplasm that are responsible for the successful improvement of crop tolerance. Later on, these genetic resources need to transfer to the subsequent cultivars. However, the CRISPR/Cas9 Genome-Editing System, phenotyping technology, genomic selection technology for molecular breeding are the few most advanced technologies to impart and enhance salt tolerance in soybean through varietal development [99, 100]. However, in different studies, soybean genotypes have been largely used to illuminate the morpho-physiological and biochemical responses under salt stress at different levels and durations (**Table 4**).

Plant tolerance to salt stress mostly depends on the capacity to regulate Na^+ or chloride (Cl^-) transportation to different parts of the plant from soil. Generally, tolerant soybean genotypes accumulate less amount of Cl^- in their leaf thus they can produce more photosynthates whereas the salt-sensitive lines accumulate more Cl^- in their leaves that reduces the net chl concentration in the plant [108, 114, 115]. Khan et al. [67] worked with 41 soybean genotypes under 100 and 150 mM NaCl stress where 7 genotypes performed as salt-tolerant under 150 mM NaCl, 21 genotypes performed as moderately tolerant under 100 mM NaCl, and rest was susceptible genotypes. Zhang et al. [101] worked on 257 genotypes with SSR markers to estimate epistatic association mapping for salt tolerance at the germination stage where 83 quantitative trait loci were detected. In addition, an experiment conducted by Luo et al. [116] with two wild soybean genotypes BB52 and N23232 collected from coastal and inland showed significant salt tolerance. In the greenhouse condition, an experiment was conducted hydroponically where 123 soybean accessions were found as salt-tolerant [117]. Salt stress highly restricts plant growth through imbalanced water status and disturbed ion uptake mechanisms. These phenomena were more focused in an experiment conducted by Shereen and Ansari [118] with four soybean cultivars viz. AGS-160, Loppa, Egyptian, ICAL-132 where ICAL-132 showed more tolerance in terms of the above-mentioned parameters. Cao et al. [102] experimented on 51 Indonesian accessions with 100 mM NaCl salt stress to speculate the salt tolerance rate, chl content (SPAD value). Among which 6 genotypes viz. Tambora, Sinyonya (late), Java 7, Sinyonya (early), Seputih Raman, and Ringgit (JP 30217) performed best as salt-tolerant genotypes. Hamayun et al. [119] conducted a screening of 69 cultivars of soybean under 100 mM salt stress for 2 weeks. The cultivars were placed in three groups (highly susceptible, susceptible, and tolerant) considering their performances under selected parameters. Salt stress remarkably affected root and shoot length and weight, photosynthesis, chl contents, transpiration rate, and nodule weight in most of the cultivars. Among 69 cultivars only 10 were finally considered as tolerant, 3 were susceptible, and the rest 56 were highly susceptible according to their overall performances. Four soybean cultivars were tested for salt tolerance under 80, 120, and 160 mM NaCl for 2 weeks. The tolerance level was estimated by leaf scorch score on a 1–9 scale (1 = no chlorosis; 9 = necrosis). The prominent differences between tolerant and sensitive cultivars were obtained at 120 mM NaCl in soil. Through this process, Williams and Clark were found as the most sensitive cultivars where the most tolerant were HBK R5525 and AG5905. In addition, leaf and root Na^+ and Cl^- concentrations were analyzed where it was observed that in the leaf of sensitive genotypes accumulated higher Na^+ than the tolerant one. At root, the opposite results were reported [108]. The presence of a higher amount of soluble protein in a plant is a good sign of

Accession name	Salt tolerance level	Tolerance traits	References
GmNcl1, GmNHX1	150 mM NaCl, 7 d	Photosynthetic rates, concentration of leaf K^+, Mg^{2+}, and P	[62]
Fengzitianandou, Baiqiu 1	100 mM NaCl, 7 d	Main root length, FW and DW of root, seedlings biomass, hypocotyls length	[101]
Java 7, Seputih Raman, Ringgit (JP 30217), Tambora, Sinyonya	100 mM NaCl, 35 d	Salt tolerance rate, leaf chl content (SPAD value), shoot DW	[102]
GmSALT3	200 mM NaCl, 18 d	Root and stem Na^+ content	[103]
PI 675847 A	200 mM NaCl	Leaf scorch scores, cell membrane stability, photosynthesis and biomass accumulation	[104]
SA 88, CX-415	150 mM NaCl, 23 d	Plant height, fresh and dry biomass of the shoot, relative leakage ratio, K^+/Na^+ ratio, proline (Pro) content	[105]
S04–05/150–114	90 mM NaCl, 8 h	Secondary structure contents of the protein isolates (α-helix, β-sheet, turn, and irregular conformations)	[106]
Clark	200 mM NaCl, 60 d	Nodule formation, leghemoglobin content, nitrogenase activity, auxin synthesis, MDA, and H_2O_2 content	[33]
En-b0–1, NILs72-T	150 mM NaCl, 21 d	Plant DW, nodulation, leaf greenness, and N uptake	[107]
HBK R5525	120 mM NaCl, 14 d	Leaf scorch score, leaf and root Na^+ and Cl^- concentrations	[108]
NILs72	120 mM NaCl, 28 d	Plant DW, photosynthetic rate, stomatal conductance, Na^+ and Cl^- content (leaf, stem, petiole, roots)	[109]
Pusa-9712, PS-1572	180 mM NaCl, 7 d	Seed germination, root and shoot length	[110]
GC840	160 mM NaCl, 15 d	Root and shoot length, FW and DW	[111]
F3	60 mM NaCl, 72 h	Level of gene and mRNA expression	[112]
Lee	150 mM NaCl, 45 d	Seed germination, shoot and root DW, and leaf mineral contents	[32]
BB52	300 mM NaCl, 7 d	Water potential, RWC, Pro and glycine betaine (GB) content, changes of ion content in young and mature seedings	[113]

Table 4.
Some of the salt-tolerant soybean genotypes as reported by researchers.

plant physiological state. Under salt stress, it triggers the plant signaling to express tolerant gene by protein upregulation and enhanced enzymatic activities [120]. Saad-Allah [121] conducted an experiment with six varieties of soybean (Crawford, G21, G22, G35, G82, and G83) under different levels of sea salt (8, and 16 dS m^{-1}) and found that the least salt-affected cultivar was G82 in terms of the highest protein content. In addition to the genotype screening method, superior characteristic gene can be identified and isolated through exploring natural allelic variation. In this way, the isolated superior gene can be inserted into the targeted better yield performing cultivars to develop salt stress tolerance. This gene-based allele-specific multiple markers genotyping could be a prospective approach in developing more salt-tolerant varieties in the future [122].

4.2 Seed priming for salt tolerance in soybean

Priming, in the early stage of germination, can stimulate the metabolic processes that result in an enhanced germination rate with uniform emergence, which is very helpful for seeds to withstand different stresses particularly abiotic stress (**Figure 3**). Various types of priming (hydropriming, osmo-priming, nutrient priming, chemical priming, bio-priming, etc.) with different agents like water, inorganic compounds, hormones, and nutrients are mostly used to improve the performance of plant morphology, physiology, leaf gas exchange, transpiration, photosynthesis and antioxidant activities [123, 124].

During the priming process, seeds are treated to enhance enzyme activity, decrease the imbibition period, increase metabolic reparation and improve germination-promoting metabolites. Additionally, primed seeds showed better osmotic adjustment by activating a cellular defense system that increases tolerance levels against abiotic stressors [125].

The advantageous effect of priming is also prominent in soybean, like many other plants. For instance, when pretreated (100 mg L^{-1} GA for 12 h) soybean seeds were sown, germination (88%), root and shoot length (16%), dry matter content of seedling (30%), vigor index (22%) and field emergence (8%) were increased compared to control. The number of pods $plant^{-1}$, seed pod^{-1}, and seed yield of pretreated seed also seen to perform better by giving a higher value at 47, 19, and 26%, respectively than untreated soybean seed [126]. A positive interaction between genotypes and priming was seen in the first pod length of soybean. Whereas, using proper priming procedure that specially adapted to cultivar, resulted in better first pod length of soybean which is desirable for the breeder [127]. Priming with polyethylene glycol (PEG) for 12 h with controlled low osmotic potential (−1.2) generated a higher final germination percentage while lowering MGT and electrical conductivity at soybean seedling compared to unprimed seed [128].

To mitigate the devastating effect of salinity in soybean, priming can be a promising measure. For example, soybean seed primed with benzyladenine resulted in higher leaf area, stomatal density, RWC at 11, 51, and 46%, respectively, and in turn improved plant growth and morphology even under high salt stress (250 mM NaCl) compared to a non-primed seed. Moreover, the number of

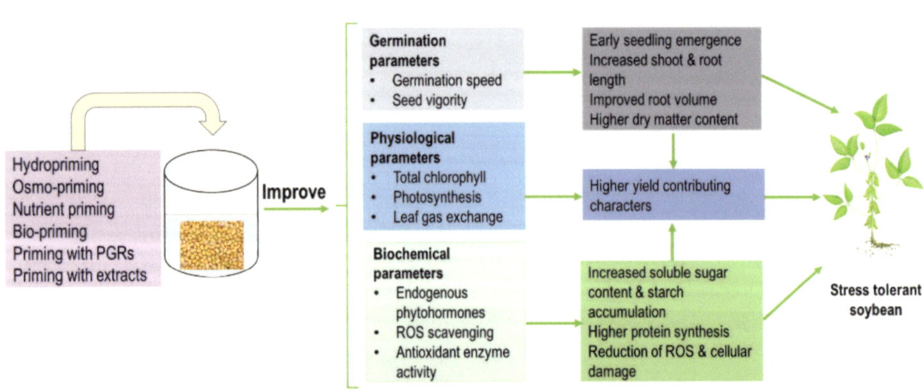

Figure 3.
Seed responses raised from primed seed lead to salt stress tolerance.

flowers and pods, which are considered as principal attributes of soybean yield, were recorded to increase by 71 and 64%, respectively in primed seed under salt stress in comparison to non-primed salt-stressed soybean. The same trend of improvement was also recorded in biochemical and physiological parameters, especially increased antioxidant activity (92%) of priming seed was noticeable [73]. Besides, improving the early growth characteristics (germination percentage, seedling shoot, and root lengths) priming can enhance the tolerance of soybean seedlings through uplifting the activities of α-amylase, protease, and nitrate reductase along with the higher $O_2^{\bullet -}$, H_2O_2, and nitric oxide (NO) contents [75]. When 5% saponin was used as a biopriming agent in soybean under salinity, shoot characteristics were seen to develop with increasing length, FW, and DW by about 15, 17, and 12%, respectively over the NaCl stressed plant (without saponin). Saponin minimized the salt-induced damages not only by ameliorating the declination of chl and carotenoid contents by about 31 and 14%, respectively, but also improving the MSI (6%) through decreasing H_2O_2 and lipid peroxidation by about 25 and 28%, respectively in comparison to stressed plant without saponin. They also made a remark that 5% saponin priming boost soybean seedling's ability to withstand salt stress by improving the antioxidant system through enhanced enzyme activities and ascorbic acid (AsA) content and decreased glutathione (GSH) [53]. A similar trend of the beneficial effect of soybean was also recorded by Sheteiwy et al. [94] by using JA as a seed priming agent with a decreased Na^+ and increased K^+ concentrations under salt stress conditions compared to non-primed seed. In addition, a balanced regulation of endogenous hormones (e.g., ABA, GA, and JA) was recorded in the case of priming that in turn enhance protein production to induce a defense mechanism under salt stress. Interestingly, priming also can take part in anatomical changes to minimize the harmful effects of salinity. Generally, salt stress causes the destruction of mesophyll cells, as plasmolysis is a common phenomenon here and consequently water absorption deficiency occurs. On the contrary, priming can elude the plasmolysis of mesophyll cells of soybean under salt stress conditions and the number and size of plastoglobuli of leaf chloroplast were also lesser than normal salt-induced seedlings, which is the sign of less damage of thylakoid and photosynthetic capacity of mesophyll cells [129]. We also observed that soybean seeds pretreated with zinc sulfate ($ZnSO_4{\cdot}7H_2O$) showed better growth performance under high salt stress (150 mM NaCl) compared to salt-stressed plants (without priming) (**Figure 4**; unpublished data). From this

Figure 4.
Effect of zinc (Zn) priming in soybean under salt stress (150 mM NaCl) (unpublished result).

evidence, it can be concluded that priming not only improves the growth and developmental process of soybean seedlings but also enriches the metabolic performance and antioxidant capacity as well under salt conditions.

4.3 Agronomic managements

Like many other crops, various agronomic management practices can enhance salt tolerance in soybean. These management practices are mainly but not limited to manipulation of sowing times, mulching, water management, tillage management, soil amendments, fertilizer management, and so on. Manipulation of agronomic practices may affect soybean tolerance to salinity because salt gradient differs among the environmental conditions such as temperature, precipitation, and humidity. For example, An et al. [130] found that elevated humidity provided soybean better tolerance to severe salinity (120 mM NaCl, 3 weeks). Elevated humidity inhibited Na^+ influx and improved root activity as well as stomatal conductance which are vital traits of salt tolerance in plants [130].

Various nutrient management showed enhanced tolerance of soybean to salt stress. In the salinity stress management of soybean, K^+ plays an active role in increasing plant growth, development, and productivity [56]. Exogenous foliar K^+ application was more efficient than seed priming in modulating soybean salt tolerance through growth, physio-biochemical attributes, yield, and seed quality. In brief, K^+ caused more than 50% improvement in shoot length, shoot DW, leaves number, leaf area in salt stressed-soybean with higher total chl and carotenoid contents. This was because of the K-induced reduction in Na^+ (69%) and Cl^- (59%) accumulation as well as higher K^+ accumulation (47%) and K^+/Na^+ ratio (361%). Higher content of total soluble sugar, Pro, and α-tocopherol with antioxidants activity by K application in stressed-plant also contributed to the suppression of ROS and membrane damage with better water content. In addition, K-mediated higher seed yield (92%) with seed protein (63%) and oil (59%) content in soybean showed its regulatory role in enhancing yield and quality attributes under salinity and proved its potentiality to be used as a sustainable approach for crop production under saline condition [56]. Calcium (Ca) is another essential plant nutrient which actively involved in plant signaling responses for increasing plant salt tolerance like soybean [131]. From the proteomic analysis, it was revealed that exogenous Ca-activated 80 and 71 proteins in cotyledon and embryo were involved in signal transduction, energy pathways, transportation, and protein biosynthesis. In addition, Ca supplementation caused the inhibition in proteolysis as well as increment in the gamma-aminobutyric acid (GABA) and polyamines, osmolytes, and secondary metabolites accumulation for attuning salt tolerance in germinating soybean [131]. Likely, as a micronutrient for instance boron (B) not only perform a significant role in plant physiology and development but also provide protection in soybean against salinity (up to 150 mM NaCl). Supplemental B regulates the different physiological processes and improves the cellular structural integrity in plants under salinity, as evidence it increased shoot FW, chl content, leaf RWC, and Pro accumulation with lower EL as growth enhancement, osmotic stress mitigation, and membrane stability indicator, respectively [92]. This elevated Pro indicated B-mediated osmoregulation to enhance osmotic status (Leaf RWC) under salinity in soybean. Thereafter, Rahman et al. [45] studied the response of soybean plants under varying degrees of saline condition with supplementation of B alone or in combination with selenium (Se). Although separate supplementation of B and Se caused better growth and water content in salt-stressed soybean, their combined foliar application was better to alleviate stress toxicity under different levels of severity. For instance, the combined application of Se and B improved leaf RWC, leaf area, shoot height, and

FW [45]. Further comprehensive studies are required to find out how and which nutrient will give more salt tolerance mechanisms together besides their nutritional value. In the study of Alharby et al. [5], only Se was applied as a foliar spray on salt-stressed soybean and confirmed Se-induced restoration of Pro, leaf RWC, higher chl accumulation. Therefore, such better growth, water status, and chl contents by 50 μM Se contributed to improvement of yield contributing and yield of soybean upon salinity. When soybean plants at an early stage were supplemented with foliar application of K (KCl and K_2SO_4), they showed enhanced tolerance to salt stress (6 and 12 dS m^{-1}) [132]. Especially, K enhanced the levels of antioxidant activities and secondary metabolites such as total polyphenol, flavonoid, chl, and carotenoid contents. However, these authors suggested that the appropriate K concentration should be fine-tuned and they also found that K_2SO_4 showed a better positive effect than KCl [132]. Organic amendments help in providing plants in improving soil and providing plant defense systems. For instance, chitosan-modified biochar (CMB) increases soybean tolerance to salt stress by enhancing plant morpho-physiology and antioxidant defense as reported by Mehmood et al. [133]. The addition of CMB resulted in 55 and 29% reduction of root Na content compared to the plants treated with salt (40 and 80 mM NaCl) alone. Shoot Na content was also decreased in the same way which was as low as 65 and 51%, respectively. These CMB treated plants produced higher osmolytes (GB and Pro), decreased H_2O_2 and MDA levels in plants. This was due to the upregulation of the antioxidant enzymes viz. catalase (CAT), ascorbate peroxidase (APX), peroxidase (POD), and superoxide dismutase (SOD) [133]. Adding water hyacinth compost and rice husk biochar (5 and 10 t ha^{-1}) also provided soybean plant tolerance to salinity (5 and 10 dS m^{-1}) as reported by Ferdous et al. [134]. However, rice husk biochar at the rate of 5 t ha^{-1} provided better results and it was prominent on the plants grown at 10 dS m^{-1} salinity.

4.4 Application of different stress elicitors

There is a necessity to utilize external stimulation in parallel with plants' inherent tolerance mechanism to tolerate salinity. Use of exogenous protectants such as phytohormones, antioxidants, amino acids, osmolytes, signaling molecules for regulating soybean salt tolerance mechanism to sustain better growth, development, and yield (**Table 5**).

Plant hormones cause plant growth and development and are widely used for that, but they also play significant roles in mitigating the adverse effects of salinity to crop production. Exogenous application of kinetin (a synthetic cytokinin) significantly improved soybean growth attributes including plant height, plant FW and DW, chl contents under saline conditions [135]. Kinetin directly enhances isoflavones biosynthesis as well as modulates phytohormone crosstalk which are involved in soybean growth regulation and stress resistance. Very recently, exogenous salicylic acid (SA) recovered significant soybean growth and yield along with elevated RWC, osmolytes accumulation, the content of chl, sugar, and starch in soybean as well as the reduction in the accumulation of toxic Na^+ under salinity [145]. Therefore, SA is able to alleviate the salt-induced both ionic and osmotic stress in soybean through regulating ionic and osmotic balance, respectively. Similarly, JA foliar spray showed better water content and water use efficiency (WUE) along with improved chl content, intercellular carbon dioxide concentration, stomatal conductance, and transpiration rate which ultimately resulted in a higher net photosynthetic rate in salt treated-soybean [94]. Therefore, JA-mediated lower Na^+ and higher K^+ accumulation, regulation of osmolyte content, and improvement in plant hormone syntheses like ABA, GA, and JA resulted in better salt tolerance mechanisms in soybean [94].

Extent of salt stress	Stress elicitors	Tolerance responses to salinity	Reference
100 mM NaCl, 14 d	5 μM kinetin	Enhanced endogenous JA and SA contents with reduction of ABA Improved growth and chl content Improved isoflavones concentration	[135]
100 mM NaCl, 23 d	5 μM GA_3	Promoted shoot length, plant FW and DW, chl content Increased daidzein and genistein contents Improved bioactive gibberellins (GA_1 and GA_4) and JA Declined endogenous ABA and SA	[136]
10 dS m^{-1} NaCl	1 mM SA	Increased N and sulfur (S) contents Improved seed yield and quality with higher content of protein and amino acid	[86]
100 mM NaCl	200 mg L^{-1} SA	Increased chl content Augmented Pro accumulation Increased sugar and starch contents	[137]
100 mmol L^{-1} NaCl, 21 d	60 μmol L^{-1} JA	Improved water content, water potential, osmotic potential, and WUE Increased net photosynthetic rate, transpiration rate, stomatal conductance, intercellular CO_2, and total chl content by 60, 51, 78, 40, and 42%, respectively Improved ABA, GA, JA by 61, 63, and 52%, respectively Decreased Na^+ concentration with higher K^+ in both leave and root tissue Increased shoot and root length with both FW and DW	[94]
80 mM NaCl, 16 d	10 μM NO as 2,2′ (hydroxynitrosohydrazono) bis-ethanimine	Improved plant growth as evidenced by higher shoot, root, and nodule weights and nodule numbers Lowered H_2O_2 to basal levels Reduced cell death	[138]
100 mM NaCl, 7 d	100 μM NO as SNP	Decreased Na^+ content with higher K^+ and Ca^{2+} uptake Increased germination percentage Increased anthocyanins and flavonoid content	[27]
70 mM NaCl, 15 d	50 μM NO as SNP	Improved shoot and root length Reduced browning, tissue drying, and increased plant survivability Increased chl and Pro contents with better RWC Stabilized Na^+ and K^+ ion ratio	[139]
100 mM NaCl, 28 d	10 μM NO as SNP	Increased root FW (38%), shoot FW (29%), whole plant DW (75%), root DW (58%), shoot DW (27%), and whole plant DW (29%) Increased chl *a* (38%), and chl *b* (44%) contents Reduced Na^+ content in leaf (117%), and root (119%) Improved K^+ content in leaf and root	[140]

Extent of salt stress	Stress elicitors	Tolerance responses to salinity	Reference
200 mM NaCl, 12 h	150 μM NO as SNP, 2 d	Increased ABA content and reduced stomatal conductance Increased RWC and chl content	[141]
15 mM NaCl	Pro, 25 mM	Lowered Na^+ accumulation. Elevated K^+ and N contents Increased shoot DW and Pro content with reduction of EL	[142]
11 dS m^{-1} NaCl	Foliar spray of GB at 10 kg ha^{-1}	Reduced Na^+ and Cl^- uptake Increased number of lateral branches (33%), pods $plant^{-1}$ (49%), and grain yield (71%) Increased endogenous GB content	[143]
50 mM NaCl, 7 d	2 mM GSH	Increased plant height, branch number Increased pod number about 12–60% with increment in seed number pod^{-1} Improved 100-seed weight as well as yield $plant^{-1}$ Reduced H_2O_2 and MDA contents	[46]
150 mM NaCl, 39 d	1 mM GSH	Increased shoot length, leaf number, and leaf area Enhanced pod number, 100-seed weight, seed protein, and oil percentage Improved total chl content, chl florescence, and performance index. Increased RWC and membrane stability with reduction of EL Elevated AsA, GSH, and α-tocopherol content Reduced Na^+ uptake with higher accumulation of N, P, K, and Ca	[58]
3 and 6 g L^{-1} NaCl, 30 d	40 mg L^{-1} cysteine	Augmented the content of chl *a*, *b*, and carotenoid Elevated Pro accumulation and the content of N, P, and K Lowered the generation of H_2O_2 and MDA Increased yield with higher seed oil content	[54]
7.46 dS m^{-1} NaCl, 70 d	6 mM K as K_2SO_4	Enhanced leaf number (74%), leaf area (52%), and shoot DW (56%) Increased chl and carotenoid contents by 185 and 20%, respectively Enhanced K^+ content with reduction of Na^+ and Cl^- uptake. Improved Pro accumulation Reduced the contents of H_2O_2 and MDA Improved seed yield and quality	[56]
6 dS m^{-1} NaCl, 45 d	50 μM Se, as Na_2SeO_4	Increased plant growth and chl content Improved Pro content with better water status. Reduced H_2O_2 generation and membrane damage Improved seed yield	[5]

Extent of salt stress	Stress elicitors	Tolerance responses to salinity	Reference
9 dS m^{-1} NaCl, 8 weeks	AsA at 100 and 200 mg L^{-1}	Increased nutrients contents like N, P, K, Fe, manganese (Mn), and Zn Reduced Pro content Improved chl and carotenoid content Improved growth, yield attributes, and quality.	[81]
Sandy coastal soil	Seaweed compost, 60 t ha^{-1}.	Increased plant growth rate, root DW, and pod number	[144]

Table 5.
Exogenous stress elicitors-induced salt tolerance in soybean.

Signaling molecules have important roles in improving various physiological attributes and adaptive mechanisms to recover plant growth against salinity. Nitric oxide (as sodium nitroprusside, SNP) supplementation significantly increased K^+ and Ca^{2+} uptake with inhibition of toxic Na^+ uptake and thus promoted ion homeostasis followed by stimulated activities of polyphenol oxidase and phenylalanine ammonia lyase of soybean in salt exposure [27]. Thus, NO-mediated higher content of flavonoids and anthocyanins suggested the NO might be affected the biosynthesis pathways of pigments to confer salinity tolerance. But the combined application of NO and SA showed more effective functional roles in salt mitigation by decreasing NaCl-induced damaging effects than individual use. The exogenous NO is also effective in alleviating the long salt toxicity of soybean as evidenced by higher growth parameters like not only shoot and root growth but also nodule weights and nodule number [138]. In addition, NO-mediated stimulation in antioxidants activities was reported which contributed to attaining higher salt tolerance and growth. From the recent study, it has been disclosed about NO-mediated higher chl content, better growth attributes, and maintenance in ion homeostasis [140]. The role of osmoprotectants including Pro and GB had been proved to increase salt tolerance in soybean with the indication of suppression in osmotic, ionic, and oxidative stress markers significantly [146].

As an important bioregulator, the amino acid has effective regulatory roles in plant growth, development, and productivity with active roles in scavenging excess ROS, thus increasing researchers' thirst to use in the regulation of plant stress management. In accordance, as a precursor of GSH containing non-protein thiol, cysteine actively improves plant stress responses upon adverse environmental conditions [54]. Foliar application of cysteine significantly improved nutrient accumulation by plants with higher osmoregulation reflected as augmented Pro content in salt-treated soybean. Not only that, cysteine-treated plants showed lower ROS generation and membrane damage as recovered from salt-induced oxidative stress. Therefore, salinity-mediated growth inhibition with suffering from lower photosynthetic pigments contents alleviated by cysteine application and thus soybean showed higher tolerance attributs under stress condition.

Glutathione is a vital component of plant antioxidant defense mechanism and also plays a key role in regulating ROS management, thus GSH had been used as a protectant to increase salt tolerance of soybean as well as to explore the mechanism of GSH [46]. Such exogenous GSH application improved soybean salt tolerance behavior through minimizing stress-induced oxidative stress and thus caused improvement in yield attributing characters leading to higher yield.

4.5 Use of beneficial microbes

Besides plant growth improvement, microbial inoculants including bacteria, fungi, and microbial symbiosis have been already enlisted for increasing plant stress tolerance including salinity [72]. Plant growth-promoting rhizobacteria (PGPR) are the potential to improve plant growth by mitigating salt toxicity, where plant and soil health are benefitted from the interaction of roots with these microorganisms and plant roots [80]. *Bacillus firmus* SW5 showed the protective role on salt stressed-soybean by increasing growth and biomass production [80]. Bacterial inoculation also caused higher nutrient accumulation, the content of chl, osmolytes (GB and Pro), soluble sugar, phenolic compound and also gas exchange parameters in stressed-soybean. Both *Bradyrhizobium japonicum* USDA 110 and *Pseudomonas putida* TSAU1 coordinately enhanced soybean growth and root architectural traits under saline conditions due to higher auxin production [147]. This better root growth in length, number, and surface area later contributed to higher nutrient absorption from the soil. Five different strains of PGPR including *Arthrobacter woluwensis* AK1, *Microbacterium oxydans*, *A. aurescens*, *B. megaterium*, and *B. aryabhattai* significantly improved soybean tolerance to salinity [4]. These salt-tolerant stains not only improved nutrient accumulation and chl biosynthesis, but also caused the improvement in hormonal regulation like higher indole-3-acetic acid (IAA), GA production with decreasing ABA content in soybean under 200 mM NaCl. In addition, halotolerant PGPR bacteria could be used as a biological safe tool for increasing plant growth by alleviating salt toxicity [4]. Arbuscular mycorrhizal fungi (AMF) were used for increasing the salinity tolerance of both salt-tolerant and sensitive cultivars of soybean [33]. Salt-stressed soybean showed better plant growth with higher nodule formation, leghemoglobin content, and nitrogenase activity under salinity with AMF inoculation irrespective of tolerance level, whereas higher AMF inoculation was detected in tolerant genotype. Plant hormones have a stimulating role in plant growth maintenance as an essential member of metabolites and in such fungal symbiosis with soybean, auxin exhibited a prime signaling role in between AMF and host plants. Hashem et al. [33] observed such increment in IAA and indole-3-butyric acid (IBA) level in AMF inoculated salt-stressed soybean regardless of cultivars. Therefore, the salt tolerance ability of soybean through microbial association is enhanced due to the stimulation in endogenous growth hormone followed by better root growth, nutrient acquisition [33, 147]. These AMF also contributes to mitigate salt-induced oxidative stress in soybean by suppressing membrane damage and ROS generation [33]. Beneficial microbes-induced modulation of different mechanisms for attaining better salt tolerance in soybean has been summarized in **Table 6**.

4.6 Enhancing antioxidant defense

Enhancing the productivity of the antioxidant defense system and synthesis of antioxidant enzymes may provide a safeguard against salinity-induced oxidative stress [154]. In soybean, salt tolerance can be elevated by enhancing the activity of antioxidant enzymes viz. CAT, APX, glutathione peroxidase (GPX), glutathione reductase (GR), POD. Higher enzymatic activity helps to repair the ROS-induced membrane dysfunctions, which ultimately accelerates plant growth by maintaining chloroplasts and other cell organelles. Such regulatory effects of the antioxidant defense system were observed in different experiments (**Table 7**). Application of different types of phytoprotectants, trace elements, nutrient elements, or organic acids play vital roles to increase plant antioxidant defense capacity under salt stress [154].

Salinity levels	Microbial inoculation	Tolerance responses to salinity	Reference
200 mM NaCl	Mycorrhizal fungi *Funneliformis mosseae*, *Rhizophagus intraradices,* and *Claroideoglomus etunicatum*	Improved the nodule formation and plant root structure Increased nutrient accumulation Enhanced the content of auxin and chl Lowered H_2O_2 production and membrane damage	[33]
100 mM NaCl, 22 d	*Bradyrhizobium japonicum*	Prompted root nodulation and seedling growth Increased chl and carotenoids content Improved maximum photochemical efficiency of PSII Decreased EL Maintained ultrastructure of thylakoid and chloroplast of mesophyll cells Raised root isoflavone content	[148]
120 mM NaCl, 10 d	*Bacillus amyloliquefaciens* H-2-5	Enhanced plant growth with 10% higher shoot length and GA_4 content Lowered the concentration of ABA, SA, JA, and Pro	[149]
150 mM NaCl	*Rhizobium* sp. SL42 and *Hydrogenophaga* sp. SL48, co-inoculated with *B. japonicum* 532C	Increased shoot biomass, seed weight, and grain yield. Enhanced N assimilation and shoot K^+/Na^+	[150]
100 mM NaCl, 7 d	*Pseudomonas* sp. strain AK- 1, and *Bacillus* sp. strain SJ-5	Increased plant FW with higher chl content Enhanced water content with higher Pro accumulation and thus lowered osmotic injury Elevated photosynthesis activity Suppressed MDA production	[14]
70 and 140 mM NaCl, 7 d	*Porostereum spadiceum* AGH786	Increased seedlings' growth and reduced transpiration rate Enhanced GA content with reduction of ABA and JA production Increased isoflavones content	[151]
120 mM NaCl, 10 d	*Pseudomonas putida* H-2-3	Enhanced the shoot length and FW Increased chl content. Reduced the contents of ABA and SA with higher content of JA Decreased Na^+ content Lowered the total polyphenol with increment in total flavonoid content	[152]
75 mM NaCl, 42 d	*B. japonicum* BDYD1 *Stenotrophomonas rhizophila* ep-17	Increased plant growth attributes like root length, shoot length, shoot and root DW Elevated nutrient uptake including N and P with a higher number of nodules	[147]
75, and 150 mM NaCl, 46 d	*Arthrocnemum macrostachyum,* 7 and 14 d	Increased shoot and root length, FW of shoot and root, and their DW with higher leaf number Reinforced the chl *a*, *b,* and carotenoids, soluble sugars, and proteins contents Decreased MDA and H_2O_2 contents Increased Pro, total free amino acids, total phenols, and AsA content	[51]

Salinity levels	Microbial inoculation	Tolerance responses to salinity	Reference
70 and 140 mM NaCl, 7 d	*Metarhizium anisopliae* pretreatment, 21 d	Increased plant growth with higher leaf area and chl content Improved transpiration and photosynthesis rate Elevated Pro content with suppression of MDA generation Reduced ABA with higher JA contents Increased isoflavonoids content from 9 to 17%	[153]

Table 6.
Beneficial microbes-mediated mechanism in increasing salt tolerance of soybean.

Dose and duration of salt stress	Changes in antioxidant defense level and enzyme activities	Reference
150 mM NaCl, 39 d	The activities of SOD, CAT, and GPX were increased Enzymatic protein content in leaf was reduced	[58]
200 mM NaCl, 21 d	FeSOD, POD, CAT, and APX activities in root were improved	[94]
3 and 6 g L^{-1} NaCl, 30 d	Increased CAT activity, but reduced SOD activity	[54]
100 mM NaCl, 25 d	Salt stress increased SOD, CAT, APX, and GR activities by 31, 16, 20, and 11%, respectively	[53]
7.46 dS m^{-1} NaCl	The activities of SOD, CAT, APX, and GPX were increased	[56]
100 mM NaCl, 2 d	CAT and APX activities were increased by 3.6 and 1.4-folds, respectively	[59]
200 and 400 mM NaCl, 69 d	Activities of CAT, APX, and GR were increased with increasing salt concentrations APX activity was higher in Clark under 200 mM salt treatment	[31]
100 mM NaCl, 30 d	SOD, APX, and GR activities were increased by 37, 40, and 33%, respectively Reduced AsA by 16% while increasing GSH and tocopherol content by 27 and 15%, respectively	[91]
100, 200, 300 mM NaCl, 10 h	Increased the activities of polyphenol oxidase and POD Total protein content was increased Reduced GSH concentration was observed	[4]

Table 7.
Regulation of antioxidant defense system by enhancing enzyme activities in soybean.

In addition, different gene expression is triggered by salt stress that can enhance the activity of antioxidant enzymes to provide a defense under prolonged salt conditions. For example, Mehmood et al. [133] observed the higher activity of four antioxidant genes (*CAT*, *APX*, *POD*, and *SOD*) and two salt-tolerance conferring genes (*GmSALT3* and *CHS*) under 40 and 80 mM salt stress in 13 soybean cultivars. The result suggested that when the CMB was used as a protectant under salt stress, the expression profile of salt-tolerant genes *GmSALT3* and *CHS* get increased and their expressions are high during the vegetative stage of the crop which is significantly affected during salinity stress.

Superoxide dismutase, as an antioxidant enzyme plays a vital role under different abiotic stresses, and their functions are well documented during oxidative stress. However, the role of *SOD* family genes under salt stress is little explored. Lu et al. [155] observed the *SOD* gene expression under 50 mm $NaHCO_3$ for 6 and 12 h. However, the soybean transcriptome data under alkaline stress revealed that six soybean *SOD* genes were differentially expressed under salt stress. Among them, only *GmFSD3*, *GmFSD5*, and *GmCSD5* were all up-regulated under alkaline and salt stress, which denote that they might have a positive regulatory role under such stress condition. The differential expression of *GmFSD3* and *GmFSD5* in soybean leaves and roots suggests that these two genes may be involved in different signaling pathways under salt stress.

5. Conclusion and perspectives

Salinity has a destructive effect on plants by imparting osmotic stress as well as ionic imbalance and toxicity. Soybean is a moderately salt-sensitive crop and cannot withstand saline conditions for a long duration. Most of the cultivated soybean are glycophytic in nature and originated through the domestication or cross-breeding with the wild type of soybean. This is why the yield performance, growth, and quality of the cultivated soybean are lower under salt stress. Soybean is a crop of versatile uses as both human food and animal feed. Therefore, the yield or prior reproductive development should be emphasized while considering different attributes for studies. Avoidance or escape mechanisms should be introduced to ensure the expected yield from soybean whether it is cultivated in favorable or unfavorable environmental conditions. Many research works have been conducted related to the soybean responses and tolerance to salt stress. However, these results are largely inconsistent due to the genotypic and experimental variations. However, the precise mechanisms of salt stress tolerance and finding the biochemical, molecular, and genetic bases of such mechanisms should be investigated comprehensively. Tailoring salt-tolerant traits is, therefore, a vital task for future plant biologists. Screening large number of genotypes and finding the appropriate genotype for saline environment would overcome the hindering of soybean production in those areas. These interventions would provide breeders and agronomists with climate-resilient soybean cultivation packages in the changing world to ensure global food security.

Acknowledgements

We acknowledge the Ministry of Science and Technology (MoST), Government of the People's Republic of Bangladesh, and Sher-e-Bangla Agricultural University Research System (SAURES), for providing funds in the research on salt stress in soybean. We thank Md. Rakib Hossain Raihan, Shamima Sultana, Ayesha Siddika, Maliha Rahman Falguni, and Khadeja Sultana Sathi for providing important literature on soybean and salt stress.

Conflict of interest

The authors declare no conflict of interest.

Author details

Mirza Hasanuzzaman[1*], Khursheda Parvin[2], Taufika Islam Anee[1], Abdul Awal Chowdhury Masud[1] and Farzana Nowroz[1]

1 Department of Agronomy, Sher-e-Bangla Agricultural University, Dhaka, Bangladesh

2 Department of Horticulture, Sher-e-Bangla Agricultural University, Dhaka, Bangladesh

*Address all correspondence to: mhzsauag@yahoo.com

References

[1] Graham PH, Vance CP. Legumes: Importance and constraints to greater use. Plant Physiology. 2003;**131**:872-877

[2] Liu S, Zhang M, Feng F, Tian Z. Toward a "green revolution" for soybean. Molecular Plant. 2020;**3**:688-697

[3] Hasanuzzaman M, Nahar K, Rahman A, Mahmud JA, Hossain MS, Fujita M. Soybean production and environmental stresses. In: Miransari M, editor. Environmental Stresses in Soybean Production. London: Academic Press; 2016. pp. 61-102

[4] Khan MA, Asaf S, Khan AL, Ullah I, Ali S, Kang SM, et al. Alleviation of salt stress response in soybean plants with the endophytic bacterial isolate *Curtobacterium* sp. SAK1. Annals of Microbiology. 2019;**69**:797-808

[5] Alharby HF, Hasanuzzaman M, Al-Zahrani HS, Hakeem KR. Exogenous selenium mitigates salt stress in soybean by improving growth, physiology, glutathione homeostasis and antioxidant defense. Phyton—International Journal of Experimental Botany. 2021;**90**: 373-388

[6] Liu Y, Yu L, Qu Y, Chen J, Liu X. GmSALT3, which confers improved soybean salt tolerance in the field, increases leaf Cl^- exclusion prior to Na^+ exclusion but does not improve early vigor under salinity. Frontiers in Plant Science. 2016;**7**:697. DOI: 10.3389/fpls.2016.01485

[7] Nahar K, Hasanuzzaman M, Alam MM, Rahman A, Suzuki T, Fujita M. Polyamines confer salt tolerance in mung bean by reducing sodium uptake, improving nutrient homeostasis, antioxidant defense and methylglyoxal detoxification systems. Frontiers in Plant Science. 2016;**7**:1104. DOI: 10.3389/fpls.2016.01104

[8] Parvin K, Nahar K, Hasanuzzaman M, Bhuyan MHMB, Mohsin SM, Fujita M. Exogenous vanillic acid enhances salt tolerance of tomato: Insight into plant antioxidant defense and glyoxalase systems. Plant Physiology and Biochemistry. 2020;**150**:109-120

[9] Perkins EG. Composition of soybeans and soybean products. In: Erickson DR, editor. Practical Handbook of Soybean Processing and Utilization. Urbana: AOCS Press; 1995. pp. 9-28

[10] Nair M, Giri RR, Boddepalli VN, Prasad KVSV, Devulapalli R, Blümmel M. Variation in grain yield, fodder quality and animal intake in two dual purpose legume crops: Mungbean and vegetable soybean grown in semi-arid tropical India Ramakrishnan. Legume Research. 2019;**44**:207-214

[11] Zhu H, Chen J, He Z, Hao W, Liu J, Kwek E, et al. Soybean germ oil reduces blood cholesterol by inhibiting cholesterol absorption and enhancing bile acid excretion. Food and Function. 2019;**10**:1836-1845

[12] Sanjukta S, Rai AK. Production of bioactive peptides during soybean fermentation and their potential health benefits. Trends in Food Science and Technology. 2016;**50**:1-10

[13] Kalenik TK, Costa R, Motkina EV, Kosenko TA, Skripko OV, Kadnikova IA. Technological development of protein-rich concentrates using soybean and meat by-products for nutrition in extreme conditions. Acta Scientiarum Polonorum, Technologia Alimentaria. 2017;**16**:255-268

[14] Kumari S, Vaishnav A, Jain S, Varma A, Choudhary DK. Bacterial-mediated induction of systemic tolerance to salinity with expression of stress alleviating enzymes in soybean (*Glycine max* L. Merrill). Journal of

Plant Growth Regulation. 2015;**34**:558-573

[15] Banaszkiewicz T. Nutritional value of soybean meal. Soybean and Nutrition. 2011;**12**:1-20

[16] Niwinska B, Witaszek K, Niedbała G, Pilarski K. Seeds of n-GM soybean varieties cultivated in Poland and their processing products as high-protein feeds in cattle nutrition. Agriculture. 2020;**10**:174. DOI: 10.3390/agriculture10050174

[17] González-Félix ML, Maldonado-Othón CA, Perez-Velazquez M. Effect of dietary lipid level and replacement of fish oil by soybean oil in compound feeds for the shortfin corvina (*Cynoscion parvipinnis*). Aquaculture. 2015;**454**:217-228

[18] Biketi A. Suitability of soybean-blood meal mixture as replacement of fish meal in nile tilapia (*Oreochromis niloticus*) rations. African Environmental Review Journal. 2020;**4**:103-110

[19] Dietzel R, Liebman M, Ewing R, Helmers M, Horton R, Jarchow M, et al. How efficiently do corn-and soybean-based cropping systems use water? A systems modeling analysis. Global Change Biology. 2016;**22**:666-681

[20] Hunt ND, Hill JD, Liebman M. Cropping system diversity effects on nutrient discharge, soil erosion, and agronomic performance. Environmental Science & Technology. 2019;**53**:1344-1352

[21] Ahmad S, Zhu X, Wei X, Zhang S. Influence of process parameters on hydrothermal modification of soybean residue: Insight into the nutrient solid biofuel, and thermal properties of hydrochars. Journal of Environmental Management. 2021;**283**:11981. DOI: 10.1016/j.jenvman.2021.111981

[22] Luo J, Zhou Y, Gao Q, Li J, Yan N. From wastes to functions: A new soybean meal and bark-based adhesive. ACS Sustainable Chemistry & Engineering. 2020;**29**:10767-10773

[23] Zhang Y, Ding J, Wang H, Su L, Zhao C. Biochar addition alleviate the negative effects of drought and salinity stress on soybean productivity and water use efficiency. BMC Plant Biology. 2020;**20**:288. DOI: 10.1186/s12870-020-02493-2

[24] Wang Z, Wang J, Bao Y, Wu Y, Zhang H. Quantitative trait loci controlling rice seed germination under salt stress. Euphytica. 2011;**178**:297-307

[25] Zhu JK. Abiotic stress signaling and responses in plants. Cell. 2016;**167**:313-324

[26] Waqas M, Korres NE, Khan MD, Nizami AS, Deeba F, Ali I, et al. Advances in the concept and methods of seed priming. In: Hasanuzzaman M, Fotopoulos V, editors. Priming and Pretreatment of Seeds and Seedlings. Singapore: Springer; 2019. pp. 11-41

[27] Simaei M, Khavari-Nejad RA, Bernard F. Exogenous application of salicylic acid and nitric oxide on the ionic contents and enzymatic activities in NaCl-stressed soybean plants. American Journal of Plant Sciences. 2012;**3**:1495-1503

[28] Shu K, Qi Y, Chen F, Meng Y, Luo X, Shuai H, et al. Salt stress represses soybean seed germination by negatively regulating GA biosynthesis while positively mediating ABA biosynthesis. Frontiers in Plant Science. 2017;**8**:1372. DOI: 10.3389/fpls.2017.01372

[29] Hosseini MK, Powell AA, Bingham IJ. Comparison of the seed germination and early seedling growth of soybean in saline conditions. Seed Science Research. 2002;**12**:165-172

[30] Xu XY, Fan R, Zheng R, Li CM, Yu DY. Proteomic analysis of seed

germination under salt stress in soybeans. Journal of Zhejiang University Science B. 2011;**12**:507-517

[31] Abulfaraj AA, Jalal RS. Use of plant growth-promoting bacteria to enhance salinity stress in soybean (*Glycine max* L.) plants. Saudi Journal of Biological Sciences. 2021;**28**: 3823-3834

[32] Essa TA. Effect of salinity stress on growth and nutrient composition of three soybean (*Glycine max* L. Merrill) cultivars. Journal of Agronomy and Crop Science. 2002;**188**:86-93

[33] Hashem A, Abd_Allah EF, Alqarawi AA, Wirth S, Egamberdieva D. Comparing symbiotic performance and physiological responses of two soybean cultivars to arbuscular mycorrhizal fungi under salt stress. Saudi Journal of Biological Sciences. 2019;**26**:38-48

[34] Khan MSA, Karim MA, Mahmud AA, Parveen S, Bazzaz MM, Hossain MA. Plant water relations and proline accumulations in soybean under salt and water stress environment. Journal of Plant Science. 2015;**3**:272-278

[35] Julkowska MM, Hoefsloot HCJ, Mol S, Feron R, De Boer GJ, Haring MA, et al. Capturing Arabidopsis root architecture dynamics with root-fit reveals diversity in responses to salinity. Plant Physiology. 2014;**166**:1387-1402

[36] Silva BRS, Batista BL, Lobato AKS. Anatomical changes in stem and root of soybean plants submitted to salt stress. Plant Biology. 2021;**23**:57-65

[37] Otie V, Udo I, Shao Y, Itam MO, Okamoto H, An P, et al. Salinity effects on morpho-physiological and yield traits of soybean (*Glycine max* L.) as mediated by foliar spray with brassinolide. Plants. 2021;**10**:541. DOI: 10.3390/plants10030541

[38] Kumari A, Parida AK. Metabolomics and network analysis reveal the potential metabolites and biological pathways involved in salinity tolerance of the halophyte *Salvadora persica*. Environmental and Experimental Botany. 2018;**148**:85-99

[39] Deinlein U, Stephan AB, Horie T, Luo W, Xu G, Schroeder JI. Plant salt-tolerance mechanisms. Trends in Plant Science. 2014;**19**:371-379

[40] Choat B, Drayton WM, Brodersen C, Matthews MA, Shackel KA, Wada H, et al. Measurement of vulnerability to water stress-induced cavitation in grapevine: A comparison of four techniques applied to a long-vesseled species. Plant, Cell & Environment. 2010;**33**:1502-1512

[41] Sellami S, Le Hir R, Thorpe MR, Aubry E, Wolff N, Vilaine F, et al. Arabidopsis natural accessions display adaptations in inflorescence growth and vascular anatomy to withstand high salinity during reproductive growth. Plants. 2019;**8**:61. DOI: 10.3390/plants8030061

[42] Amirjani MR. Effect of salinity stress on growth, mineral composition, proline content, antioxidant enzymes of soybean. American Journal of Plant Physiology. 2010;**5**:350-360

[43] Weisany W, Sohrabi Y, Heidari G, Siosemardeh A, Ghassemi-Golezani K. Changes in antioxidant enzymes activity and plant performance by salinity stress and zinc application in soybean (*Glycine max* L.). Plant Omics. 2012;**5**:60-67

[44] Lee SK, Sohn EY, Hamayun M, Yoon JY, Lee IJ. Effect of silicon on growth and salinity stress of soybean plant grown under hydroponic system. Agroforestry Systems. 2010;**80**:333-340

[45] Rahman M, Rahman K, Sathi KS, Alam MM, Nahar K, Fujita M, et al. Supplemental selenium and boron mitigate salt-induced oxidative damages

in *Glycine max* L. Plants. 2021;**10**:2224. DOI: 10.3390/plants10102224

[46] Akram S, Siddiqui MN, Hussain BMN, Bari MAA, Mostofa MG, Hossain MA, et al. Exogenous glutathione modulates salinity tolerance of soybean [*Glycine max* (L.) Merrill] at reproductive stage. Journal of Plant Growth Regulation. 2017;**36**:877-888

[47] Egbichi I, Keyster M, Jacobs A, Klein A, Ludidi N. Modulation of antioxidant enzyme activities and metabolites ratios by nitric oxide in short-term salt stressed soybean root nodules. South African Journal of Botany. 2013;**88**:326-333

[48] Roy SJ, Negrão S, Tester M. Salt resistant crop plants. Current Opinion in Biotechnology. 2014;**26**:115-124

[49] Munns R. Comparative physiology of salt and water stress. Plant, Cell & Environment. 2002;**25**:239-250

[50] Sheteawi SA. Improving growth and yield of salt-stressed soybean by exogenous application of jasmonic acid and ascobin. International Journal of Agriculture and Biology. 2007;**9**:473-478

[51] Osman MS, Badawy AA, Osman AI, Abdel Latef AAH. Ameliorative impact of an extract of the halophyte *Arthrocnemum macrostachyum* on growth and biochemical parameters of soybean under salinity stress. Journal of Plant Growth Regulation. 2021;**40**: 1245-1256

[52] Liu A, Xiao Z, Li MW, Wong FL, Yung WS, Ku YS, et al. Transcriptomic reprogramming in soybean seedlings under salt stress. Plant, Cell & Environment. 2019;**42**:98-114

[53] Soliman MH, Abdulmajeed AM, Alhaithloul H, Alharbi BM, El-Esawi MA, Hasanuzzaman M, et al. Saponin biopriming positively stimulates antioxidants defense, osmolytes metabolism and ionic status to confer salt stress tolerance in soybean. Acta Physiologiae Plantarum. 2020;**42**:114. DOI: 10.1007/s11738-020-03098-w

[54] Sadak MS, Abd El-Hameid AR, Zaki FS, Dawood MG, El-Awadi ME. Physiological and biochemical responses of soybean (*Glycine max* L.) to cysteine application under sea salt stress. Bulletin of the National Research Centre. 2020;**44**:1. DOI: 10.1186/s42269-019-0259-7

[55] Ghassemi-Golezani K, Farhangi-Abriz S, Bandehagh A. Salicylic acid and jasmonic acid alter physiological performance, assimilate mobilization and seed filling of soybean under salt stress. Acta Agriculturae Slovenica. 2018;**111**:597-607

[56] Taha RS, Seleiman MF, Alotaibi M, Alhammad BA, Rady MM, Mahdi AHA. Exogenous potassium treatments elevate salt tolerance and performances of *Glycine max* L. by boosting antioxidant defense system under actual saline field conditions. Agronomy. 2020;**10**:1741. DOI: 10.3390/agronomy10111741

[57] Farhangi-Abriz S, Torabian S. Biochar improved nodulation and nitrogen metabolism of soybean under salt stress. Symbiosis. 2018;**74**:215-223

[58] Zaki SS, Belal EEE, Rady MM. Cyanobacteria and glutathione applications improve productivity, nutrient contents, and antioxidant systems of salt-stressed soybean plant. International Letters of Natural Sciences. 2019;**76**:72-85

[59] Chung YS, Kim KS, Hamayun M, Kim Y. Silicon confers soybean resistance to salinity stress through regulation of reactive oxygen and reactive nitrogen species. Frontiers in Plant Science. 2020;**10**:1725. DOI: 10.3389/fpls.2019.01725

[60] Yan K, He W, Bian L, Zhang Z, Tang X, An M, et al. Salt adaptability in a halophytic soybean (*Glycine soja*) involves photosystems coordination. BMC Plant Biology. 2020;**20**:155. DOI: 10.1186/s12870-020-02371-x

[61] Fang S, Hou X, Liang X. Response mechanisms of plants under saline-alkali stress. Frontiers in Plant Sciense. 2021;**12**:667458. DOI: 10.3389/fpls.2021.667458

[62] Ning L, Kan G, Shao H, Yu D. Physiological and transcriptional responses to salt stress in salt-tolerant and salt-sensitive soybean (*Glycine max* [L.] Merr.) seedlings. Land Degradation and Development. 2018;**29**:2707-2719

[63] Abdel-Hafeez AN, El-Mageed TA, Rady MM. Impact of ascorbic acid foliar spray and seed treatment with cyanobacteria on growth and yield component of sunflower plants under saline soil conditions. International Letters of Natural Sciences. 2019;**76**:136-146

[64] Latef AA, Alhmad MF, Kordrostami M, Abo-Baker AB, Zakir A. Inoculation with *Azospirillum lipoferum* or *Azotobacter chroococcum* reinforces maize growth by improving physiological activities under saline conditions. Journal of Plant Growth Regulation. 2020;**39**:1293-1306

[65] Guo R, Yang Z, Li F, Yan C, Zhong X, Liu Q, et al. Comparative metabolic responses and adaptive strategies of wheat (*Triticum aestivum*) to salt and alkali stress. BMC Plant Biology. 2015;**15**:170. DOI: 10.1186/s12870-015-0546-x

[66] Katerji N, Van Hoorn JW, Hamdy A, Mastrorilli M. Salt tolerance classification of crops according to soil salinity and to water stress day index. Agricultural Water Management. 2000;**43**:99-109

[67] Khan F, Hakeem KR, Siddiqi TO, Ahmad A. RAPD markers associated with salt tolerance in soybean genotypes under salt stress. Applied Biochemistry and Biotechnology. 2013;**170**:257-272

[68] Dadkhah AR, Grrifiths H. The effect of salinity on growth, inorganic ions and dry matter partitioning in sugar beet cultivars. Journal of Agricultural Science and Technology. 2006;**8**:199-210

[69] Ferdous J, Mannan MA, Haque MM, Mamun MA, Alam MS. Chlorophyll content, water relation traits and mineral ions accumulation in soybean as influenced by organic amendments under salinity stress. Australian Journal of Crop Science. 2018;**12**:1806-1812

[70] Mannan MA, Karim MA, Haque MM, Khaliq QA, Higuchi H, Nawata E. Response of soybean to salinity: III. Water status and accumulation of mineral ions. Tropical Agriculture and Development. 2013;**57**:41-48

[71] Benedetti L, Scherner A, Cuchiara C, C, Moraes ÍL, A Avila L, Deuner S. Soybean plant osmotic and oxidative stress as affected by herbicide and salinity levels in soil. Planta Daninha. 2020;**38**:e020185196. DOI: 10.1590/S0100-83582020380100517

[72] Hasanuzzaman M, Raihan MRH, Masud AAC, Rahman K, Nowroz F, Rahman M, et al. Regulation of reactive oxygen species and antioxidant defense in plants under salinity. International Journal of Molecular Sciences. 2021;**22**:9326. DOI: 10.3390/ijms22179326

[73] Mangena P. Role of benzyladenine seed priming on growth and physiological and biochemical response of soybean plants grown under high salinity stress condition. International Journal of Agronomy. 2020;**5**:8847098. DOI: 10.1155/2020/8847098

[74] Ghassemi-Golezani K, Taifeh-Noori M, Oustan S, Moghaddam M,

Seyyed-Rahmani S. Oil and protein accumulation in soybean grains under salinity stress. Notulae Scientia Biologicae. 2010;**13**:64-67

[75] Kataria S, Baghel L, Jaina M, Guruprasad KN. Magnetopriming regulates antioxidant defense system in soybean against salt stress. Biocatalysis and Agricultural Biotechnology. 2019;**18**:101090. DOI: 10.1016/j.bcab.2019.101090

[76] Baghel L, Kataria S, Guruprasad KN. Static magnetic field treatment of seeds improves carbon and nitrogen metabolism under salinity stress in soybean. Bioelectromagnetics. 2016;**37**:455-470

[77] Farhangi-Abriz S, Ghassemi-Golezani K. How can salicylic acid and jasmonic acid mitigate salt toxicity in soybean plants? Ecotoxicology and Environmental safety. 2018;**147**: 1010-1016

[78] Hamayun M, Khan SA, Khan AL, Shinwari ZK, Hussain J, Sohn EY, et al. Effect of salt stress on growth attributes and endogenous growth hormones of soybean cultivar Hwangkeumkong. Pakistan Journal of Botany. 2010;**42**:3103-3112

[79] Khan MA, Asaf S, Khan AL, Adhikari A, Jan R, Ali S, et al. Halotolerant rhizobacterial strains mitigate the adverse effects of NaCl stress in soybean seedlings. BioMed Research International. 2019;**2019**: 953963. DOI: 10.1155/2019/9530963

[80] El-Esawi MA, Alaraidh IA, Alsahli AA, Alamri SA, Ali HM, Alayafi AA. *Bacillus firmus* (SW5) augments salt tolerance in soybean (*Glycine max* L.) by modulating root system architecture, antioxidant defense systems and stress-responsive genes expression. Plant Physiology and Biochemistry. 2018;**132**:375-384

[81] Abdelhamid MS, Gaballah MS, Rady M, Gomaa A. Biofertilizer and ascorbic acid alleviated the detrimental effects of soil salinity on growth and yield of soybean. In: Proceedings of The Second Science with Africa Conference. Addis Ababa. 2010. pp. 73-81

[82] Parveen AU, Akhtar J, Basra SM. Interactive effect of salinity and potassium on growth, biochemical parameters, protein and oil quality of soybean genotypes. Pakistan Journal of Agricultural Sciences. 2016;**53**:69-78

[83] Khan MSA, Karim MA, Haque MM, Islam MM, Karim AJMS, Mian MAK. Influence of salt and water stress on growth and yield of soybean genotypes. Pertanika Journals Tropical Agricultural Science. 2016;**39**:167-180

[84] Oyebowale AM, Moses EI, Ndiko L. Comparative analysis of responses to field salinity stress in contrasting soybean accessions highlights NaCl exclusion in leaves as a key mechanism for salinity stress tolerance. Proceedings of the 4th Tunisian-South African Mini-symposium on Biodiversity in Forage and Feed Crops for Sustainable Productivity under Environmental Constraints. Journal of Oasis Agriculture and Sustainable Development. 2021;(special issue):19-24

[85] Ghassemi-Golezani K, Farhangi-Abriz S. Foliar sprays of salicylic acid and jasmonic acid stimulate H^+-ATPase activity of tonoplast, nutrient uptake and salt tolerance of soybean. Ecotoxicology and Environmental Safety. 2018;**166**:18-25

[86] Farhangi-Abriz S, Ghassemi-Golezani K. Improving amino acid composition of soybean under salt stress by salicylic acid and jasmonic acid. Journal of Applied Botany and Food Quality. 2016;**89**:243-248

[87] Parveen A-U-HM, Aziz T, Aziz O, Maqsood L. Potassium induces carbohydrates accumulation by enhancing morpho-physiological and

biochemical attributes in soybean under salinity. Archives of Agronomy and Soil Science. 2020;**67**:946-959

[88] Hasanuzzaman M, Nahar K, Rahman A, Anee TI, Alam MU, Bhuiyan TF, et al. Approaches to enhance salt stress tolerance in wheat. In: Wanyera R, editor. Wheat Improvement, Management and Utilization. Rijeka: Intech; 2017. pp. 151-187

[89] Alam I, Sharmin SA, Kim K-H, Kim Y-G, Bahk JD, Lee B-H. Comparative proteomic approach to identify proteins involved in flooding combined with salinity stress in soybean. Plant and Soil. 2011;**346**:45-62

[90] Liu HR, Sun GW, Dong LJ, Yang LQ, Yu SN, Zhang SL, et al. Physiological and molecular responses to drought and salinity in soybean. Biologia Plantarum. 2017;**61**:557-564

[91] Elkelish AA, Alnusaire TS, Soliman MH, Gowayed S, Senousy HH, Fahad S. Calcium availability regulates antioxidant system, physio-biochemical activities and alleviates salinity stress mediated oxidative damage in soybean seedlings. Journal of Applied Botany and Food Quality. 2019;**92**:258-266

[92] Alharby HF, Nahar K, Al-Zahrani HS, Hakeem KR, Hasanuzzaman M. Enhancing salt tolerance in soybean by exogenous boron: Intrinsic study of the ascorbate-glutathione and glyoxalase pathways. Plants. 2021;**10**:2085. DOI: 10.3390/plants10102085

[93] Alharbi BM, Elhakem AH, Alnusairi GSH, Mona H, Soliman MH, Hakeem KR, et al. Exogenous application of melatonin alleviates salt stress-induced decline in growth and photosynthesis in *Glycine max* (L.) seedlings by improving mineral uptake, antioxidant and glyoxalase system. Plant, Soil and Environment. 2021;**67**: 208-220

[94] Sheteiwy MS, Shao H, Qi W, Daly P, Sharma A, Shaghaleh H, et al. Seed priming and foliar application with jasmonic acid enhance salinity stress tolerance of soybean (*Glycine max* L.) seedlings. Journal of the Science of Food and Agriculture. 2020;**101**:2027-2041

[95] Klein A, Keyster M, Ludidi N. Response of soybean nodules to exogenously applied caffeic acid during NaCl-induced salinity. South African Journal of Botany. 2015;**96**:13-18

[96] Hasanuzzaman M, Nahar K, Fujita M. Plant response to salt stress and role of exogenous protectants to mitigate salt-induced damages. In: Ahmad P, Azooz MM, Prasad MNV, editors. Ecophysiology and Responses of Plants under Salt Stress. New York: Springer; 2013. pp. 25-87

[97] Munns R, James RA, Läuchli A. Approaches to increasing the salt tolerance of wheat and other cereals. Journal of Experimental Botany. 2006;**57**:1025-1043

[98] Cao D, Li Y, Liu B, Kong F, Tran LS. Adaptive mechanisms of soybean grown on salt-affected soils. Land Degradation & Development. 2018;**29**:1054-1064

[99] Chen H, Liu XQ, Zhang HM, Yuan XX, Gu HP, Cui XY, et al. Advances in salinity tolerance of soybean: Genetic diversity, heredity, and gene identification contribute to improving salinity tolerance. Journal of Integrative Agriculture. 2018;**17**:2215-2221

[100] Chen H, Heng Y, Tuyen DD, Zhou J, Valliyodan B, Shannon GJ, et al. Advances in genetics and breeding of salt tolerance in soybean. In: Kumar V, Wani S, Suprasanna P, Tran LS, editors. Salinity Responses and Tolerance in Plants. Vol. 2. Cham: Springer; 2018. pp. 217-237

[101] Zhang WJ, Niu Y, Bu SH, Li M, Feng JY, Zhang J, et al. Epistatic

association mapping for alkaline and salinity tolerance traits in the soybean germination stage. PLoS One. 2014;**9**:e84750. DOI: 10.1371/journal.pone.0084750

[102] Cao D, Yan YL, Xu DH. Assessment of salt tolerance and analysis of the salt tolerance gene NaCl in Indonesian soybean germplasm. Plant Genetic Resources. 2019;**17**:265-271

[103] Guan R, Qu Y, Guo Y, Yu L, Liu Y, Jiang J, et al. Salinity tolerance in soybean is modulated by natural variation in G m SALT 3. The Plant Journal. 2014;**80**:937-950

[104] Quach TN, Nguyen LA, Nguyen NT, Nguyen CT, Dam TQ, Dinh LT, et al. Soybean PI 675847 A as a new source of salt tolerance. Plant Genetic Resources. 2019;**17**:33-44

[105] Cicek N, Çakırlar H. Effects of salt stress on some physiological and photosynthetic parameters at three different temperatures in six soybean (*Glycine max* L. Merr.) cultivars. Journal of Agronomy and Crop Science. 2008;**194**:34-46

[106] Akyuz S, Akyuz T, Celik O, Atak C. FTIR spectroscopy of protein isolates of salt-tolerant soybean mutants. Journal of Applied Spectroscopy. 2018;**84**:1019-1023

[107] Song Y, Nakajima T, Xu D, Homma K, Kokubun M. Genotypic variation in salinity tolerance and its association with nodulation and nitrogen uptake in soybean. Plant Production Science. 2017;**20**:490-498

[108] Ledesma F, Lopez C, Ortiz D, Chen P, Korth KL, Ishibashi T, et al. A simple greenhouse method for screening salt tolerance in soybean. Crop Science. 2016;**56**:585-594

[109] Onodera M, Nakajima T, Nanzyo M, Takahashi T, Xu D, Homma K, et al. Regulation of root-to-leaf Na and Cl transport and its association with photosynthetic activity in salt-tolerant soybean genotypes. Plant Production Science. 2019;**22**:262-274

[110] Kumari S, Lal SK, Sreenivasa V, Rajendran A, Singh SK, Singh KP, et al. Screening and identification of sources of salinity tolerance at seed germination stage in Indian soybean genotypes. International Journal of Current Microbiology and Applied Sciences. 2019;**8**:3006-3013

[111] Moniruzzaman M, Saiem RM, Emon RM, Haque MS, Saha NR, Malek MA, et al. Genetic diversity analysis of soybean genotypes using SSR markers for salinity tolerance. Progressive Agriculture. 2019;**30**:1-9

[112] Rahmawati N, Rosmayati D, Basyuni M, Oku H. Evaluation of resistance improvement of soybean (*Glycine max* (L) Merr.) against salinity using mass selection and gene expression of salinity tolerant. In: Proceedings of The International Conference on Natural Resources and Technology (ICONART). Medan. 2019. pp. 30-37

[113] Wu G, Zhou Z, Chen P, Tang X, Shao H, Wang H. Comparative ecophysiological study of salt stress for wild and cultivated soybean species from the Yellow River Delta, China. The Scientific World Journal. 2014. DOI: 10.1155/2014/651745

[114] Patil G, Do T, Vuong TD, Valliyodan B, Lee JD, Chaudhary J, et al. Genomic-assisted haplotype analysis and the development of high-throughput SNP markers for salinity tolerance in soybean. Scientific Reports. 2016;**6**:19199. DOI: 10.1038/srep19199

[115] Zeng A, Chen P, Korth K, Hancock F, Pereira A, Brye K, et al. Genome-wide association study (GWAS) of salt tolerance in worldwide soybean germplasm lines. Molecular Breeding. 2017;**37**(3):1-4

[116] Luo Q, Yu B, Liu Y. Differential sensitivity to chloride and sodium ions in seedlings of *Glycine max* and *G. soja* under NaCl stress. Journal of Plant Physiology. 2005;**162**:1003-1012

[117] Do TD, Chen H, Hien VTT, Hamwieh A, Yamada T, Sato T, et al. Ncl synchronously regulates Na^+, K^+, and Cl^- in soybean and greatly increases the grain yield in saline field conditions. Scientific Reports. 2016;**6**:19147. DOI: 10.1038/srep19147

[118] Shereen A, Ansari R, Soomro AQ. Salt tolerance in soybean (*Glycine max* L.): Effect on growth and ion relations. Pakistan Journal of Botany. 2001;**33**: 393-402

[119] Hamayun M, Khan SA, Iqbal A, Hussain A, Lee IJ. Screening of soybean cultivars for salinity tolerance under hydroponic conditions. Fresenius Environmental Bulletin. 2019;**28**: 7955-7963

[120] Kosova K, Prášil IT, Vítámvás P. Protein contribution to plant salinity response and tolerance acquisition. International Journal of Molecular Sciences. 2013;**14**:6757-6789

[121] Saad-Allah KM. Impact of sea salt stress on growth and some physiological attributes of some soybean (*Glycine max* L.) varieties. Iranian Journal of Plant Physiology. 2015;**6**:1559-1571

[122] Kumawat G, Maranna S, Gupta S, Tripathi R, Agrawal N, Singh V, et al. Identification of novel genetic sources for agronomic and quality traits in soybean using multi-trait allele specific genic marker assays. Journal of Plant Biochemistry and Biotechnology. 2021;**30**:160-171

[123] Paparella S, Araújo SS, Rossi G, Wijayasinghe M, Carbonera D, Balestrazzi A. Seed priming: State of the art and new perspectives. Plant Cell Reports. 2015;**34**:1281-1293

[124] Rehman A, Mazhar K, Ijaz M, Ali Q, Ahmad S. Seedling pretreatment: Methods and protocols. In: Hasanuzzaman M, Fotopoulos V, editors. Priming and Pretreatment of Seeds and Seedlings. Singapore: Springer; 2019. pp. 117-134

[125] Hussain S, Hussain S, Khaliq A, Ali S, Khan I. Physiological, biochemical, and molecular aspects of seed priming. In: Hasanuzzaman M, Fotopoulos V, editors. Priming and Pretreatment of Seeds and Seedlings. Singapore: Springer; 2019. pp. 43-62

[126] Agawane RB, Parhe SD. Effect of seed priming on crop growth and seed yield of soybean [*Glycine max* (L.) MERILL]. The Bioscan. 2015;**10**: 265-270

[127] Lewandowska S, Łoziński M, Marczewski K, Kozak M, Schmidtke K. Influence of priming on germination, development, and yield of soybean varieties. Open Agriculture. 2020;**5**: 930-935

[128] Sadeghi H, Khazaei F, Yari L, Sheidaei S. Effect of seed osmopriming on seed germination behavior and vigor of soybean (*Glycine max* L.). Journal of Agricultural and Biological Sciences. 2011;**6**:39-43

[129] Dai LY, Zhu HD, Yin KD, Du JD, Zhang YX. Seed priming mitigates the effects of saline-alkali stress in soybean seedlings. Chilean Journal of Agricultural Research. 2017;**77**:118-125

[130] An P, Inanaga S, Kafkafi U, Lux A, Sugimoto Y. Different effect of humidity on growth and salt tolerance of two soybean cultivars. Biologia Plantarum. 2001;**44**:405-410

[131] Yin Y, Yang R, Han Y, Gu Z. Comparative proteomic and physiological analyses reveal the protective effect of exogenous calcium on the germinating soybean response to

salt stress. Journal of Proteomics. 2015;**113**:110-126

[132] Adhikari B, Dhungana SK, Kim ID, Shin DH. Effect of foliar application of potassium fertilizers on soybean plants under salinity stress. Journal of the Saudi Society of Agricultural Sciences. 2020;**19**:261-269

[133] Mehmood S, Ahmed W, Ikram M, Imtiaz M, Mahmood S, Tu S, et al. Chitosan modified biochar increases soybean (*Glycine max* L.) resistance to salt-stress by augmenting root morphology, antioxidant defense mechanisms and the expression of stress-responsive genes. Plants. 2020;**9**:1173. DOI: 10.3390/plants9091173

[134] Ferdous J, Mannan MA, Haque MM, Alam MS, Talukder S. Mitigation of salinity stress in soybean using organic amendments. Bangladesh Agronomy Journal. 2018;**21**:39-50

[135] Hamayun M, Hussain A, Khan SA, Irshad M, Khan AL, Waqas M, et al. Kinetin modulates physio-hormonal attributes and isoflavone contents of soybean grown under salinity stress. Frontiers in Plant Science. 2015;**6**:377. DOI: 10.3389/fpls.2015.00377

[136] Hamayun M, Khan SA, Khan AL, Shin JH, Ahmad B, Shin DH, et al. Exogenous gibberellic acid reprograms soybean to higher growth and salt stress tolerance. Journal of Agricultural and Food Chemistry. 2010;**58**:7226-7232

[137] Jaiswal A, Pandurangam V, Sharma SK. Effect of salicylic acid in soybean (*Glycine max* L. Meril) under salinity stress. The Bioscan. 2014;**9**:671-676

[138] Egbichi I, Keyster M, Ludidi N. Effect of exogenous application of nitric oxide on salt stress responses of soybean. South African Journal of Botany. 2014;**90**:131-136

[139] Karthik S, Pavan G, Krishnan V, Sathish S, Manickavasagam M. Sodium nitroprusside enhances regeneration and alleviates salinity stress in soybean [*Glycine max* (L.) Merrill]. Biocatalysis and Agricultural Biotechnology. 2019;**19**:101173. DOI: 10.1016/j.bcab.2019.101173

[140] Jabeen Z, Fayyaz HA, Irshad F, Hussain N, Hassan MN, Li J, et al. Sodium nitroprusside application improves morphological and physiological attributes of soybean (*Glycine max* L.) under salinity stress. PLoS One. 2021;**16**:e0248207. DOI: 10.1371/journal.pone.0248207

[141] Dinler BS, Antoniou C, Fotopoulos V. Interplay between GST and nitric oxide in the early response of soybean (*Glycine max* L.) plants to salinity stress. Journal of Plant Physiology. 2014;**171**:1740-1747

[142] Sabagh AE, Islam MS, Ueda A, Saneoka H, Barutçular C. Increasing reproductive stage tolerance to salinity stress in soybean. International Journal of Agriculture and Crop Sciences. 2015;**8**:738-745

[143] Rezaei MA, Kaviani B, Masouleh AK. The effect of exogenous glycine betaine on yield of soybean [*Glycine max* (L.) Merr.] in two contrasting cultivars Pershing and DPX under soil salinity stress. Plant Omics. 2012;**5**:87-93

[144] Purwaningsih O, Pamungkas PB, Beny D, Oktavia M. Response of soybean growth in sandy coastal soil to seaweed compost and biochar application. Journal of Physics: Conference Series. 2021;**1823**:012028. DOI: 10.1088/1742-6596/1823/1/012028

[145] Nigam B, Dubey RS, Rathore D. Protective role of exogenously supplied salicylic acid and PGPB (*Stenotrophomonas* sp.) on spinach and soybean cultivars grown under salt

stress. Scientia Horticulturae. 2022;**293**:110654. DOI: 10.1016/j.scienta.2021.110654

[146] El-Sabagh A, Sorour S, Omar AE, Ragab A, Islam MS, Barutçular C, et al. Alleviation of adverse effects of salt stress on soybean (*Glycine max*. L.) by using osmoprotectants and organic nutrients. International Journal of Agricultural and Biosystems Engineering. 2015;**9**:1014-1018

[147] Egamberdieva D, Jabborova D, Berg G. Synergistic interactions between *Bradyrhizobium japonicum* and the endophyte *Stenotrophomonas rhizophila* and their effects on growth, and nodulation of soybean under salt stress. Plant and Soil. 2016;**405**:35-45

[148] Meng N, Yu BJ, Guo JS. Ameliorative effects of inoculation with *Bradyrhizobium japonicum* on *Glycine max* and *Glycine soja* seedlings under salt stress. Plant Growth Regulation. 2016;**80**:137-147

[149] Kim MJ, Radhakrishnan R, Kang SM, You YH, Jeong EJ, Kim JG, et al. Plant growth promoting effect of *Bacillus amyloliquefaciens* H-2-5 on crop plants and influence on physiological changes in soybean under soil salinity. Physiology and Molecular Biology of Plants. 2017;**23**:571-580

[150] Ilangumaran G, Schwinghamer TD, Smith DL. Rhizobacteria from root nodules of an indigenous legume enhance salinity stress tolerance in soybean. Frontiers in Sustainable Food Systems. 2021;**4**:308. DOI: 10.3389/fsufs.2020.617978

[151] Hamayun M, Hussain A, Khan SA, Kim HY, Khan AL, Waqas M, et al. Gibberellins producing endophytic fungus *Porostereum spadiceum* AGH786 rescues growth of salt affected soybean. Frontiers in Microbiology. 2017;**8**:686. DOI: 10.3389/fmicb.2017.00686

[152] Kang SM, Radhakrishnan R, Khan AL, Kim MJ, Park JM, Kim BR, et al. Gibberellin secreting Rhizobacterium, Pseudomonas putida H-2-3 modulates the hormonal and stress physiology of soybean to improve the plant growth under saline and drought conditions. Plant Physiology and Biochemistry. 2014;**84**:115-124

[153] Khan AL, Hamayun M, Khan SA, Kang SM, Shinwari ZK, Kamran M, et al. Pure culture of *Metarhizium anisopliae* LHL07 reprograms soybean to higher growth and mitigates salt stress. World Journal of Microbiology and Biotechnology. 2012;**28**:1483-1494

[154] Phang TH, Shao G, Lam HM. Salt tolerance in soybean. Journal of Integrative Plant Biology. 2008;**50**:1196-1212

[155] Lu W, Duanmu H, Qiao Y, Jin X, Yu Y, Yu L, et al. Genome-wide identification and characterization of the soybean SOD family during alkaline stress. Peer J. 2020;**8**:e8457. DOI: 10.7717/peerj.8457

Chapter 4

Accelerated Methods of Determining Wheat Genotypes Primary Resistance to Extreme Temperatures

Alexandru Dascaliuc

Abstract

Several morphological and functional mechanisms determine the resistance of plants to extreme temperatures. Depending on the specificity of mechanisms of action, we divided them into two groups: (1) the mechanisms that ensure the avoidance/reduction of the exposure dose; (2) functional mechanisms which increase plant resistance and ability to recover damages caused by stress through regulation metabolic and genes expression activity. We developed theoretical and practical methods to appreciate the contribution of parameters from both groups on the primary and adaptive resistance of different wheat genotypes. This problem became more complicated because some properties are epigenetically inherited and can influence genotypes' primary (initial) resistance to stressors. The article describes results obtained by the accelerated determination of the initial resistance of wheat (*Triticum aestivum* L.) genotypes to temperature stress and the prospects for their implementation in the selection and development of methods for rational choosing wheat varieties for cultivation under specific environmental conditions.

Keywords: wheat genotypes, temperature stress, primary resistance, adaptive resistance, accelerated stress resistance tests

1. Introduction

As a result of the global warming trend, environmental conditions are changing; the frequency of periods with extreme temperature and their intensity is increasing. Between 1990 and 2015 the earth's global average surface temperature has linearly risen relative to the average surface temperature of 1961–1990 by 0.9°C [1]. At the same time, according to expectations, if the average temperature rises by one degree, the expected wheat production will decrease by 6% [2]. Environmental changes influence the rate of growth, development, survival, and productivity of plants. That determines the necessity to develop effective breeding methods and technologies for growing hybrids and genotypes following the specific conditions of each plant cultivation region. The success of solving these problems depends on elucidating the mechanisms that influence the resistance of plants to the action of extreme temperatures. Their knowledge allows specialists to select plant varieties well adapted to the new conditions and rationally distribute the existing genotypes for cultivation in the appropriate areas, thus achieving their biological potential.

However, depending on the duration of exposition, the temperatures outside the optimal range, specific to each genotype, can have serious consequences, causing a decrease or complete loss of the harvest. The resistance of plants to the action of temperature stress factors depends on integrating processes at the molecular, subcellular, cellular, tissular, and whole plant levels [3]. The dynamics and integration of these processes determine the primary resistance and the capacity of plants to adapt to changes in environmental conditions [3–5]. The primary (initial) resistance and adaptive capacity of plants to stressors vary [6, 7]. The morphology of plant organs [4], seed vigor, and viability [8] change under environmental conditions. They influence both the mechanisms of avoiding/reducing the dose of the stress factor, determined by multiplying the temperature value (the intensive factor) to the duration of exposure (the extensive factor), and the efficiency of the genetic and epigenetic mechanisms of inheriting the primary and adaptive plants resistance to the action of the stress factor [3, 9]. We should consider that some epigenetically inherited characteristics can be gradually eliminated [9, 10]; therefore, they do not necessarily remain inherited in the next generation. These mechanisms may cause the changes in primary and adaptive resistance of genotypes to stressors.

Given the complexity of the mechanisms that determine the resistance of plants to temperature stress and the specific manifestation in biological systems with different levels of organization and stage of ontogenesis [3, 4], it becomes clear the need to elucidate their interactions. To clarify the specificity of the response of autumn wheat plants to extreme positive and negative temperatures, we introduced in research the seeds of different wheat genotypes reproduced in Moldova and Ukraine. The survival and productivity of winter wheat plants depend on the complex processes of their adaptations to winter frosts, summer heat, and drought. Therefore, the specificity of the resistance inheritance and developing the adaptation processes to environmental conditions of wheat during winter and summer is of theoretical and practical interest. Furthermore, the reaction of biosystems to thermal stress complexly depends on the exposure dose. With increasing the value of stress-inducing temperature, the biological effect of the particular dose also increases [3, 4]. Therefore, to compare the resistance of different wheat varieties to extreme temperatures, we chose the temperature level and duration of exposure, which applied to varieties with medium resistance will cause a 50% decrease in the value of the studied parameter. The parameters can be different: percentage of seed germination, accumulated plant biomass, crop yield, etc. By applying this procedure, we were able to arrange wheat genotypes according to their resistance to stress. The objectives of our researches with different winter wheat genotypes were the following:

1. To develop the accelerated methods of determining the wheat genotypes' primary resistance to the action of high temperatures and frost.

2. To use the elaborated methods for evaluation of the wheat genotypes' primary resistance to the action of high temperatures and frost.

3. To determine the influence of the environmental conditions of seeds reproduction and their storage during 1 year on wheat genotypes' primary resistance to heat or frost.

2. Materials and methods

Our studies used the seeds of different varieties and lines of hexaploid wheat reproduced in Ukraine and Moldova. In the beginning, we calibrated the wheat seeds by volume, passing them through sieves with 2.4–2.6 mm diameter. Next, they were immersed in a 0.1% potassium permanganate solution for 20 min, thoroughly washed with a tap, then with distilled water. After soaking in water at +4°C, we exposed the seeds to heat shock (HS) by immersing them in the water at a specified temperature for a particular time, maintained with an accuracy of ±0.05°C using an ultrathermostat U10 (Germany). The shock with negative temperatures (SNT) was provided by seeds incubation in an air thermostat Rumed 3401 (Germany) for 8 h, maintaining the air temperature with an accuracy of ±0.5°C. In the studies, we installed two control variants: first for experiments with determining the wheat genotypes' resistance to frost, and second, for determining their resistance to high temperatures. Before applying for germination, we incubated the seeds of control variants at a temperature of +24°C for an additional 8 h or 30 min, equal to the duration of exposure to SNT or HS in experimental variants. Seeds of experimental and controls variants were then germinated in Petri dishes, 25 wheat seeds each, in triplicate, in the dark, at 25°C, and 75–85% relative humidity. The reaction of seeds of individual wheat genotypes to SNT, or HS, was judged by the percentage of seeds germinating within 5 days. After 5 days of incubation at +25°C, we mentioned that germinated at least 95% of wheat seeds from the control variants. Therefore, the genotype's response to SNT, or HS, was judged directly based on the percentage of seeds germinated in the experimental variants. To determine the statistical validity of the obtained data, we calculated the average value from at least three experiments repetitions, the standard deviations of average, and the credibility of the differences between the average values [11].

3. Results and discussions

3.1 Preparation of seeds for the appreciation of hexaploid wheat genotypes' primary resistance to high temperatures and frost

The data presented in **Figure 1** shows the kinetics of water accumulation by the seeds of wheat variety Odesskaya 267. We described the dynamics of increasing seeds humidity with the duration of immersion in water from 0 to 130 h by the polynomial equation of dependence $Y_x = -0.0034x^2 + 0.7114x + 12.605$, where Y_x is the moisture content of seeds depending on the duration of incubation-x. In general, there was a tendency to decrease water accumulation speed with increasing duration of immersing the seeds in water. By the rate of water accumulation, we separated the imbibition into two phases: phase I—from 0 to 30 h (initial active water adsorption), and phase II—from 30 to 130 h (plateau phase). **Figure 1** shows the data corresponding to these phases marked in blue and dark blue, respectively. The increasing of the seed's humidity in interior of phase I is described by equation $Y_x = -0.0191x^2 + 1.3766x + 7.4948$, and phase II—by equation $Y_x = -0.002x^2 + 0.4822x + 19.328$. Comparing the three equations, we can observe that in all the increases of the humidity percentage with the increase of the imbibition time, the linear component influence dominates due to the high value of the coefficient. As expected, the value of the linear coefficient in phase I of imbibition is much higher than in phase II (1.3766 and 0.4822; the ratio is equal to 2.886). For the total period of immersing in the water, in a period between 0 and 130 h, as expected, this coefficient of linear proportionality was lower than that characteristic for interval I and higher than in interval II, being equal to 0.7114. The average values of the seed moisture growth rate in each phase confirm these data. The mean growth rate of seed humidity during the total immersion time in the water was equal to 0.29% per hour when phases I and II were equal to 0.77 and 0.16% per hour.

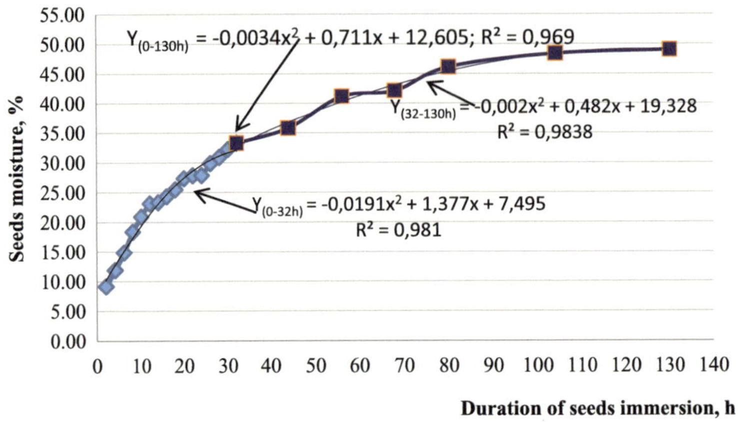

Figure 1.
Dynamics of increasing the humidity (I) in the wheat seeds, variety Odesskaya 286, depending on the duration of immersion in water at a temperature of 4°C. $Y_{(0-32h)}$, $Y_{(32-130h)}$, *and* $Y_{(0-130h)}$—*The polynomial approximation of increasing seeds humidity approximation on intervals of 0–32, 32–130, and 0–130 h of imbibition, respectively.*

During 130 h of imbibition at 4°C, the germination of the seeds did not occur. Therefore, there was no transition to faze III, characterized by a further increase in the rapidity of water imbibition and prominence of the radicle [8, 12]. The seeds' imbibition in water during 130 h at 4°C was not associated with finalizing the embryonic axis extension up to a level that ensures germination expression. The mentioned legitimacies correspond to the stages of water accumulation in the seeds in a state of dormancy described in the specialized literature. The germination and the appearance of the radicle take place only in phase III of germination [4, 9]. The radicle's formation by seeds was not detected even after 130 h of immersion in water in our experiments.

The conditions developed for preparing seeds for germination and determination of the integral kinetics of wheat seeds imbibition with water, shown in **Figure 1**, were the necessary steps for appreciating the wheat genotypes' primary resistance to high temperatures and frosts. The resistance of biological systems to biotic and abiotic stressors depends on the processes at different levels of their organization. In general, external factors can induce stress, specifically influenced by the nature, dose of the stressor, and specificity of the mechanisms determining the plant's resistance. The higher the exposition dose that initiates the stress state, the higher the biological system's resistance to a specific stressor. Stress tolerance can also be induced by external factors and allow organisms to survive the action of lethal doses when previously exposed to moderate levels of the stressor [3, 4]. This type of induced high level of extreme temperatures tolerance is maintained only for a limited period. It corresponds to the processes defined as acclimation [4]. From the above, it results that at any stage of ontogenesis, the total resistance of the plant depends on two components: (1) the initial (primary) resistance, and (2) the secondary resistance [13], induced by environmental factors, superimposed on the plant's primary resistance.

During ontogenesis, due to adaptation processes, the resistance of genotypes to stress factors varies considerably. The induced level of adaptation depends on both the genotype and the dose of stress factor [12–14]. Therefore, the correct distribution of genotypes by resistance requires their synchronization in an identical physiological state. In each moment, the entire plant's stress resistance depends on the primary stress resistance, superimposed by the induced adaptations, named the secondary resistance, and the mechanisms that assure stress avoidance [4, 12, 14]. Thus, the plants' total resistance to abiotic stress depends on three groups of mechanisms: primary resistance, resistance by avoidance, and resistance due to adaptations performed in ontogenesis. Considering this, when preparing wheat seeds for germination, the results of which are presented in **Figure 1**, the seeds imbibition with water were provided under conditions excluding the induction of adaptations. The use of seeds prepared for germination in this way opens up the possibility of directly assessing the primary resistance of the wheat genotype to extreme temperatures, excluding the influence of adaptation and avoidance mechanisms. Because of the variability of adaptive resistance, determined functionally and by avoidance, the correct determination of the genotype primary resistance is an important and challenging problem. The physiological state of plants depends not only on the genetic identity, age, and current environmental conditions but also on their previous adaptations to the environment during ontogenesis. The specially designated experiments showed that the hexaploid wheat genotypes' primary frost resistance constitutes only 14–33% of the value of total frost resistance achieved after inducing the maximum level of hardening [15, 16].

In natural conditions, the synchronization of seeds' physiological state occurs during their formation, maturation, and entry into the dormancy [8]. In the quiescent embryos of wheat seeds, cells are in the G_1 phase of the cell cycle, the transcription and the mitotic cycle are stopped [8, 13, 16, 17]. At the initiation of germination, in all seeds, the physiological states of quiescent embryos are practically identical. However, several factors can disrupt the synchronization state of embryo cells in the transition of wheat seeds from a quiescent state to the germination stage. The period of different grains maturation in the wheat plant's ear is non-uniform. In addition, the environmental conditions which can be different during the plants ripening influencing the size, structural and functional characteristics of the grains. The specificity of internal and environmental conditions during seed maturation and emergence from dormancy can disrupt the synchronization of embryonic cells due to their capture at different stages of mitosis [8, 13, 17]. Probably, the accumulation of most cells in the G_1 phase of the cell cycle manifests itself only under conditions optimal for their maturation and desiccation of the grain, when the moisture content of the embryos gradually decreases to about 10%. These factors can generally disrupt the uniformity of the release time of wheat seeds from dormancy during the germination period. Germination starts with moistening and is completed when the extending radicle penetrates the structures that surround the seed [12].

The heterogeneity of the seeds' physiological state can also increase when creating favorable conditions for germination. In this period, the metabolic processes are restored; the expression of the genes necessary for germination is gradually activated. Germination of the wheat quiescent embryo is primarily influenced by temperature and seed moisture. If exposure to temperature can be uniform and precisely controlled, then the uniform increase of different seeds moisture and readiness for germination is much more challenging to ensure. The seeds' structural-morphological differences influence the dynamics of water absorption, which determine the heterogeneity of the time of reaching the moisture level, sufficient for germination initiation. The seed germination begins with the activation of metabolic processes. In wheat embryos, protein biosynthesis initiates 30 min after immersion in water [16]. The proteins synthesized in the first 9 h after imbibition are involved in the activation of the DNA replication. Still, the final activation of DNA biosynthesis is performed only 6–8 h later, accompanied by the transition of cells from the G_1 phase to phase S (phase of DNA synthesis) of the mitotic cycle [17]. The G_1 phase of the mitotic cycle is like the gateway through which cells transit from the non-proliferative (quiescent) to the proliferative period, accompanied by the synthesis of a limited number of proteins functionally involved in the cell cycle. The biosynthesis of DNA is induced or activated. These processes precede the initiation of cell divisions that occur after the radicle protrude tegument. Activation of the cell cycle in wheat embryos begins with the induction of mitosis in the cells of root apex cells then gradually spreads to all meristematic cells [17]. The start of phase S and G_2 (post-synthesis phase) of the cycle in wheat radicle meristematic cells is observed 12–14 h after germination initiation [16]. Thus, under favorable conditions, the germination of wheat embryos will occur with 24 h *lag* period. Typically, the uniformity of the dynamics of seeds imbibition with water can substantially influence the heterogeneity of their germination. As demonstrated experimentally, at the initiation of wheat seed germination, replication of DNA triggered by proteins translated from mRNA stored in mature seeds are not sufficient for the cells to commence all stages of mitosis [16, 17].

The need to synchronize the physiological state of seeds in the pre-germination stage is due to many reasons. First, the penetration at this stage of water into the seeds causes temporary structural perturbations of membranes, which lead to a leakage of solutes and low molecular weight metabolites into the surrounding water for imbibition. Second, the diminution of the processes of metabolites leakage occurs after a transition of the membrane phospholipid components from the gel phase achieved during maturation drying to the hydrated liquid-crystalline state, after which the low molecular weight metabolites leakage is reduced [18]. A rational way to reduce the heterogeneity of seed germination time, influenced by different water absorption rates in seeds, was achieved by extending the duration of seed saturation with water at low temperatures, **Figure 1**. Under these conditions, the rate of wheat seeds saturation with water became high, but germination processes slowed down substantially. As a result of the long duration of the wheat seeds immersion in water, the reparation of structures damaged during their maturation and rehydration were completed, and the uniformity of seeds' germination increased. Data presented in **Figure 1** shows that during 130 h of water imbibition in wheat seeds at +4°C, their transition to faze III of germination deed did not occur. We noticed that after the pre-imbibition under the mentioned above conditions and the subsequent exposure of the seeds in optimal germination conditions, the seeds were characterized with the highest speed and the highest percentage of germination when immersed in water with temperature + 4°C, during 36 h. At optimal temperature, 90, 100, and 80% of seeds were germinated, previously submerged in water with +4°C, for 12, 36, and 72 h. Mass germination of seeds (more than 50% of the wheat seeds) occurred 24, 12, and 36 h after the transition to optimal conditions for germination. Thus, the immersion of the seeds in water for 12 h was too short of reaching the level of moisture and physiological transformations necessary for the rapid germination of the wheat seeds. On the contrary, more than 72 h of immersion in water was too long-lasting and inhibited seed germination (probably due to anoxia caused by excessive oxygen consumption during prolonged immersion). So, the wheat seeds previously incubated in water at a temperature of +4°C for 36 h were the best prepared for germination. Therefore, for further research on the accelerated assessment of the primary resistance of wheat genotypes to high temperatures and frost, we used seeds prepared for synchronized germination by their immersion in water for 36 h, at temperature + 4°C.

Considering that the difference of frost resistance between wheat varieties increases with the adaptation processes' accomplishments, the traditional methods of determining frost resistance ordinarily are performed after induction of adaptive resistance at the maximal level [15, 16, 19]. Although the ability to differentiate genotypes in this state by frost resistance appears to be the greatest, the widespread use of the method is difficult. The long duration of adaptation processes fulfillment and the dependence of the kinetics of adaptation from the level of genotype resistance to frost [15, 19], are among the main arising problems. Furthermore, the complexity of mentioned processes determines the low productivity of the noted method, high costs of testing, and the uncertainty of the results that depend on the influence of a large complex of factors. In particular, studies have shown that the maximum difference between the primary frost resistance of wheat varieties does not exceed 3°C, and after their complete hardening reaches 12°C [15, 16]. Despite this, the relative simplicity of the procedures providing uniform preparation for germination of seeds of different varieties before determining the primary resistance of wheat varieties to frost inspired us to assess the possibility of separating wheat genotypes by their resistance to extreme temperatures (positive and negative) in this identical state. In our experiments, in the control variant, a similar physiological state was confirmed by the practically 100% seeds germinating during the first 24 h of incubation in favorable conditions. Therefore, after reaching this equivalent physiological state, differences in the response of seeds to HS or SNT will mainly depend on genetic differences between genotypes.

3.2 Determining the wheat genotypes primary resistance to high temperatures and frost

In multiple experiments, we determined that the exposition of the moderately resistant to high temperatures or frost seeds of wheat variety Odesskayia 267 to HS at temperature 50°C during 30 min, or SNT at a temperature -7°C during 16 h, causes diminution of their germination by 50%. Based on these results, we proposed to arrange the wheat genotypes by their primary resistance to extreme temperatures by their germination reaction after seeds exposure to the mentioned dose of HS or SNT. In **Figure 2**, as an example, we provide the results that support the correct choice of HS dose caused by exposing wheat seeds to a temperature of +50°C over 30 min as one effective to separate genotypes by their primary resistance to high temperature rapidly. Analyzing the data presented in **Figure 2**, we notice that after exposure to the mentioned dose of HS, the distribution according to the percentage of germinated seeds, genotypes were uniformly occupying the area between 11.5% (variety Vdala) and 98.5% (variety Chevalier). Of the 20 wheat varieties included in the analysis, 9—showed low, 5—medium, and 6—high thermotolerance. After HS exposure, the seeds of genotypes constituting groups with low, medium, and high thermotolerance, respectively germinated up to 50%, between 50 and 70%, and more than 70%.

The data on the primary resistance to SNT or HS of the seeds of 10 wheat genotypes multiplied in Ukraine or Moldova are presented in **Figure 3**. We considered that the genotypes, whose seeds after exposure to SNT or HS germinated up to 50%, between 50 and 70%, and more than 70%, can be divided into groups with low, medium, or high resistance SNT or HS. The data presented in **Figure 3** shows that the seeds of genotypes 466, 542, 111, 1108, Samurai, and Arctis, after cultivation in Ukraine, were characterized with low resistance; those of the genotypes 517 and 1087—with medium resistance, and seeds of genotypes 21 and Toulouse—showed high resistance to SNT action. After multiplication in Moldova, the seeds of all genotypes, except those of the variety Toulouse, with medium resistance to SNT, have shown low resistance to SNT. In contrast, except for seeds of lines 21, 542, and 111 with a low resistance to HS, the multiplied in Moldova seeds of the other seven genotypes showed medium or high resistance to HS. Among them, the seeds of lines 466 and 1087, varieties Samurai and Toulouse, obtained from the plants grown in Moldova, showed relatively high resistance to HS. After cultivation in Moldova,

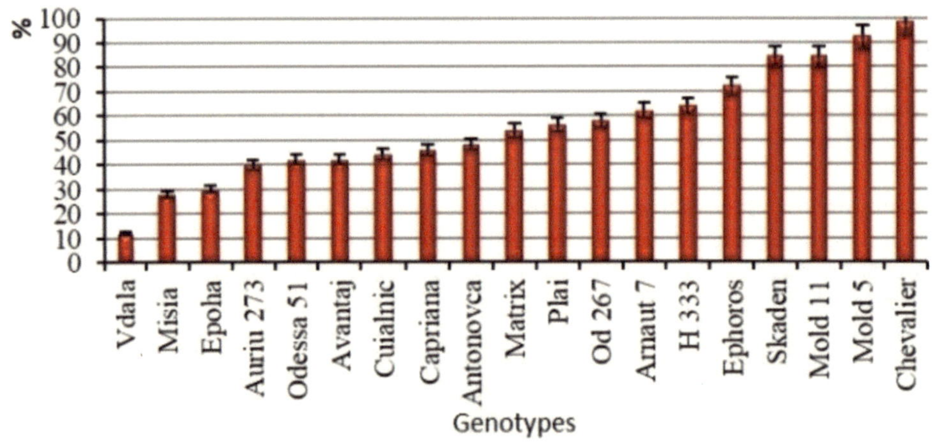

Figure 2.
Influence of seed exposure of different wheat varieties to HS at 50°C for 30 min on germination percentage over 5 days.

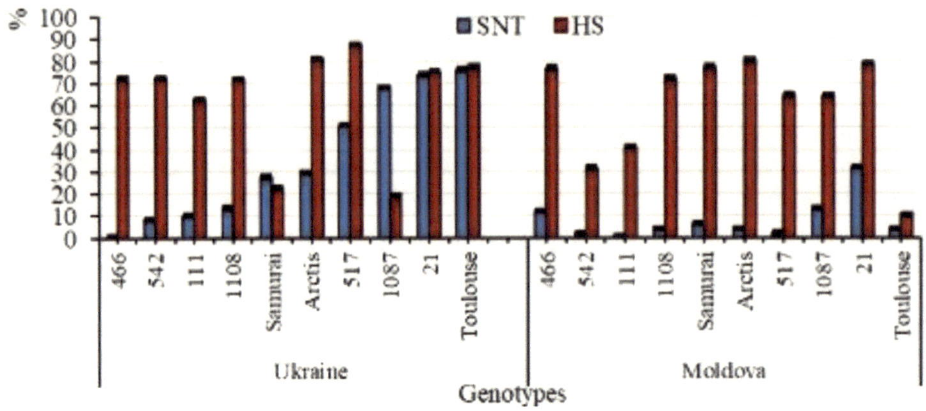

Figure 3.
The percentage of seeds of the 10 winter wheat genotypes germinated after their exposure for 16 h SNT at -7°C (blue), and that of the percentage of germinated after exposure for 30 min to HS with 50°C (red). On the left are presented the data obtained with seeds multiplied in 2015–2016 in the Kharkiv region (Ukraine), and on the right—Those obtained with the seeds multiplied in 2016–2017 in the Chisinau area (Moldova).

the correlation coefficient between the resistance of the genotypes seeds to HS and SNT reached a value of 0.434. We mention that the correlation coefficient of the resistance to SNT of the seeds obtained from the plants cultivated in Moldova and Ukraine was equal to 0.481 when the respective correlation of the seed's resistance to HS was negative to -0.314. Thus, the resistance to SNT of seeds obtained from plants grown in Moldova was lower than those multiplied in Ukraine. Nevertheless, the positive values of the correlation coefficients of resistance to SNT of seeds grown in Ukraine and Moldova suggest that a tendency towards a decrease in resistance to SNT of propagated in Moldova seeds were manifested concomitantly in most genotypes. The negative correlation coefficient between the HS resistance of the seeds multiplied in Ukraine and Moldova suggests the opposite direction of changing the HS resistance of seeds obtained from plants grown in Moldova compared to those of the seeds multiplied in Ukraine. For example, the seeds HS resistance of the lines 542, 111, and the variety Toulouse was higher for propagated in Ukraine seeds, when the resistance of the seed of the genotypes Samurai, Arctis, and 1087, on the contrary, was higher in the case of their multiplication in Moldova.

3.3 Influence of the conditions of reproduction on the plasticity of wheat genotypes resistance to extreme temperatures

The plasticity of genotype response to extreme temperatures can be characterized by their primary resistance to high temperatures and negative ones. Theoretically, all possible values of the sum obtained for seeds of the different genotypes exposed to SNT and HS occupy the range between 0 and 200%. Therefore, we considered that this sum is the limit of the maximum value of the sum of seeds percentage that will germinate after exposure to HS or SNT. By comparing the values characterizing the wheat genotypes in the experimental variants, we can conclude about the relative plasticity of genotypes resistance to extreme temperatures: the lower the value of this sum, the lower is genotype plasticity in response to extreme temperatures.

The results of the experimental determination of 10 winter wheat lines and varieties' plasticity are included in **Figure 4**. Thus, in specially designed experiments, based on the normalized response of wheat seeds to SNT or HS, we estimated each genotype's plasticity in response to the action of extreme temperatures. Based on the percentage of seed germinated after SNT or HS, we divided the wheat genotypes into three groups: genotypes with low, medium, and high plasticity. They included genotypes characterized by the value of sum less than 100%, between 100% and 140%, and higher than 140%, **Figure 4**.

Let us analyze the level of plasticity of the response to SNT and HS of the wheat genotypes seeds reproduced in Ukraine. We can mention that high plasticity was specific to seeds of the variety Toulouse, medium—those of lines 21 and 517; the seeds of other genotypes have low plasticity, **Figure 4**. The seeds of the Toulouse variety obtained from plants cultivated in Moldova have medium plasticity. In contrast, the plasticity of reaction to excessive temperatures of the seeds of nine other wheat genotypes was low. It is interesting to note that, as a rule, the higher

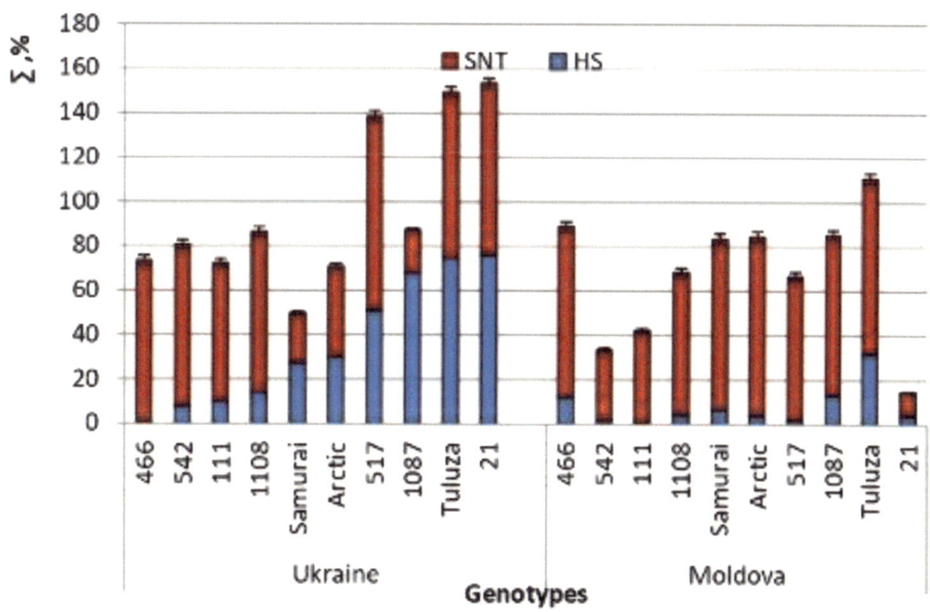

Figure 4.
The sum of the percentage of the 10 winter wheat genotypes seeds germinated after exposure to SNT for 16 h SNT at −7°C plus that of germination percentage after seeds exposure for 30 min to HS at 50°C. On the left are presented the data obtained with seeds multiplied in 2015–2016 in the Kharkiv region (Ukraine), and on the right are those obtained with seeds multiplied in 2016–2017 in the Chisinau area (Moldova).

plasticity level of multiplied in Ukraine seeds was due to their higher resistance to SNT compared to the seeds obtained from the plants cultivated in Moldova. At the same time, in general, the HS resistance of seeds reproduced in Moldova tended to be higher. After reproduction in Ukraine, only the seeds of line 21 showed high plasticity and high resistance to SNT and HS. After reproduction in Moldova, they were characterized with a low resistance to SNT, HS, and plasticity.

The original goal of our researches was to elaborate one the accelerated method for determining the primary resistance of wheat genotypes to extreme positive and negative temperatures. Hexaploid wheat cultivars are characterized by adaptation to a broad range of environmental conditions. The trials in artificial conditions with well-organized systems and testing by cultivation in different regions were accomplished [20, 21]. Hexaploid wheat cultivars respond to environmental stressors by adaptations to a large temporal seasonal variability across regions and years. The wheat cultivation in different ecological conditions indicates genetic systems that allow genotypes to cope with rapid shifts of the temperature values. The contemporary breeding programs are mainly based on the extensive and expansive testing of cultivars in different environments or climatic chambers with well-controlled temperature, illumination, and nutrition conditions. These researches demonstrated that the combined effects of multiples genes that influence morphology, growth, and development of plants, in combination, can assure the survival of wheat cultivars in conditions with extreme temperatures. Under the exposition to extreme temperatures, the adaptive changes in general plant traits: organ dimensions, shape, anatomical structure, photoperiod, vernalization, etc., are observed. It is known that enhancing wheat productivity in the Green Revolution was mainly based on the selection of semi-dwarf, insensitive to photoperiod, and disease-resistant genotypes [22]. In fluctuating environmental conditions, we must implement agrotechnical procedures that can ensure the harvest's economically convenient quantity and quality. Mechanisms of stress avoidance formed the basis of the principles that influenced the improvement of productivity of wheat plants genotypes that actually are cultivated in stressful conditions. They have influenced plants' resistance to temperature stress by changing the morphological and anatomical structure of plants. Among these changes, we note the development of the root system, penetrating more profound and more moist soil layers [20, 21], the coating of leaves with a layer of wax, the composition of pigments, the angle and shape of the leaves, which increase the refraction of light and the efficiency of transpiration [21]. The complexity of the influence of the mechanisms mentioned above of avoiding stressors caused a diversion of the attention from the genotypically determining physiological and biochemical processes, influencing the response of plants to stressors. We have made efforts to appreciate the involvement of these mechanisms in achieving by different wheat genotypes the primary resistance to extreme temperatures by exposing them to shock with positive or negative temperatures. We determined the primary resistance of wheat genotypes to the action of high temperatures using seeds well prepared for germination but which have not yet completed it. By excluding the influence of avoidance mechanisms and ontogenetic adaptations to extreme temperatures stress, we assured the possibility of comparing the primary resistance of different wheat genotypes. However, the data presented in **Figures 2–4** shows that, for the well-prepared for germination seeds, the value of the differences between the resistance of genotype to the action of extreme temperatures is relatively low compared to those obtained after induction of plant adaptation under long-time exposition to optimal for adaptation temperatures [15, 16]. Nevertheless, our experiments can differentiate the 10 wheat genotypes by their primary resistance to extreme positive or negative temperatures.

The ratio between the percentage of seeds germinated after HS and SNT in genotypes with high summary resistance to both types of shock showed a general tendency of approaching the value equal to one. Let us compare the trend of changing the values of this balance for the 10 genotypes reproduced in Ukraine and Moldova, **Figures 3** and **4**. We notice that for seeds multiplied in Moldova, the value of the primary resistance to HS reported to that of resistance to SNT, in general, is higher than this ratio evaluated for seeds reproduced in Ukraine. The shift in this ratio was primarily caused by declining the resistance to SNT of the seeds multiplied in Moldova. After reproduction in Moldova, only the variety Toulouse fell into the category of genotypes with medium summary resistance to extreme temperatures, and the others demonstrated low resistance. When reproduced in Ukraine, seeds of line 517 showed medium summary resistance, those of the Toulouse variety and the line 21—high resistance. Unexpected were the results obtained with seeds of line 21 reproduced in Moldova. If, after breeding in Ukraine, they concomitantly showed the increased resistance to HS and SNT, then after producing in Moldova, on the contrary, they were susceptible to both HS and SNT.

The data presented in this article results from efforts to assess the differences between wheat genotypes' primary resistance to extremes temperatures in the accelerated mode. Because the total resistance is the sum of the primary and adaptive resistance at any stage of ontogenesis, their discrimination is an important task. Therefore, we assumed that the summary seeds' resistance to stressors before germination is practically equivalent to the primary one. With this in mind, we have developed a method for synchronizing the physiological state of wheat seeds before the start of germination. Subsequently, using as a model the seeds of the variety Odesskaya 267, characterized by medium resistance to high temperatures and frost, we evaluated the optimal doses of HS or SNT exposure that caused the decreasing seeds germination up to 50%. The mentioned results allowed us to consider that these doses are optimal for the experimental distribution of wheat genotypes according to their primary resistance to high temperatures or frost. Furthermore, the obtained results demonstrated that seed resistance to both types of shock is specific for different genotypes but variable for the seeds of the same genotype but reproduced in other geographical conditions.

We must emphasize that the distribution of genotypes according to their primary resistance to extreme temperatures based on seeds' exposure to a single dose is simplifying. The resistance of biological systems to stressors can be multilaterally characterized by determining the doses of the stressor that causes decreasing the values of the investigated parameter by 16, 50, and 84% compared to that characteristic value for the control variant. After applying different doses of stress factors, some differences can be observed in the distribution of genotypes according to their resistance to extreme temperatures. Among the factors influencing these changes, we note the influence of seeds priming with inorganic salts, plant water extracts, organic molecules [23], and biostimulators [24, 25] on wheat stress resistance and productivity and as well the next generation inherited epigenetically acquired characters [26]. Sometimes these traits are inherited from one generation to the next. Therefore, we cannot exclude that the level of primary resistance, determined by the proposed method, is not influenced by epigenetic inheritance. The latter can be variable in the level of expression and duration of heritage. The practical techniques of elucidating epigenetic processes' integral role in installing plant adaptations to environmental conditions are under development. The preliminary data obtained show that the level of shock dose with extreme temperatures influences the distribution of wheat genotypes according to their primary resistance, which may be a consequence of the specific development of stress-initiated processes of genetic adaptations and simultaneously be influenced by the epigenetic inheritance.

We assume that due to the overlapping influence of genetic and epigenetic factors on the wheat varieties' stress resistance, their descendants are characterized by variable primary resistance to HS and SNT and high productivity in very different environmental conditions. Compared to those reproduced in Ukraine, the data obtained with the seeds multiplied in Moldova confirm this suggestion. Genotypes, the seeds of which in our experiments showed essential differences in primary resistance to high or negative temperatures in the case of reproduction in Ukraine or Moldova, are of particular interest to assess the possible influence of the epigenetic inheritance on their primary and adaptive resistance to extreme temperatures depending on the conditions of reproduction.

3.4 Modification of the plasticity of wheat genotypes primary resistance to extreme temperatures during seeds storage

Phenotypic and genotypic analysis has shown that the longevity of wheat seeds is variable, involves different regions of the genome, and is not associated with the genes that determine the state of the seeds' dormancy [27]. Among the genes that influence longevity seeds are those that control the morphological structure of the spike or the response to abiotic and biotic stressors [28]. As the duration of seeds storage increases, their viability [29] and resistance to stress factors decrease [30]. Related to this appears the question of whether the mentioned changes affect the primary resistance of different wheat genotypes seeds. To elucidate this question, we included in the research the seeds of different varieties of hexaploid wheat collected in 2019 in the Chisinau area and kept in the dark for 1 year in a thermostat, at a temperature of 20°C and relative air humidity 67–80%. Throughout the storage period, the seed moisture deed does not exceed 10%. **Figure 5** includes data on the results of determining the germination percentage of seeds of different wheat genotypes after exposure to SNT. The data were obtained for two variants of seeds: the seeds on which the exposure to SNT was carried out 2 months after harvesting and seeds stored for 1 year before testing. They show that after 1 year of storage, the primary resistance of seeds showed the tendency to decrease for all varieties. The decreasing level varied between 0%, for seeds of the variety Moldova 79 and 3.5%, for the seeds of variety Moldova 66. Although the correlation coefficient between the level of primary frost resistance after 1 year of seeds storage and the resistance before that period was very high, being equal to 0.996, this does not mean that the changes in primary resistance were equivalent at all varieties. This high value of the correlation coefficient is since the storage of seeds of all varieties during 1 year did not significantly influence the absolute value of the percentage of seeds that germinated after exposure to SNT. The correctness of this hypothesis can be confirmed by the fact that the correlation coefficient between the percentage of seeds germinated after exposure to SNT immediately after harvest and the values of the differences between the rate of seeds that grow directly after harvest and a year later was much lower, being equal only 0.190.

Figure 6 includes data on the results of determining the germination percentage of seeds of different wheat genotypes after exposure to HS. From a qualitative point of view, the data on the influence of the different wheat genotypes seeds storage during 1 year on the resistance to HS, shown in **Figure 6**, are similar to those that characterize the impact of seed storage on their resistance SNT, **Figure 5**. After exposure to HS, seed storage during 1 year caused a relatively moderate decrease in germination percentage (minimum value was 3.8% for seeds of the variety Pisanca, and the maximum value was 6.3% for the seeds of variety Moldova 66). The correlation coefficient between the germination percentage of seeds exposed to HS immediately after harvesting (in 2019) and those after 1 year of seeds storage

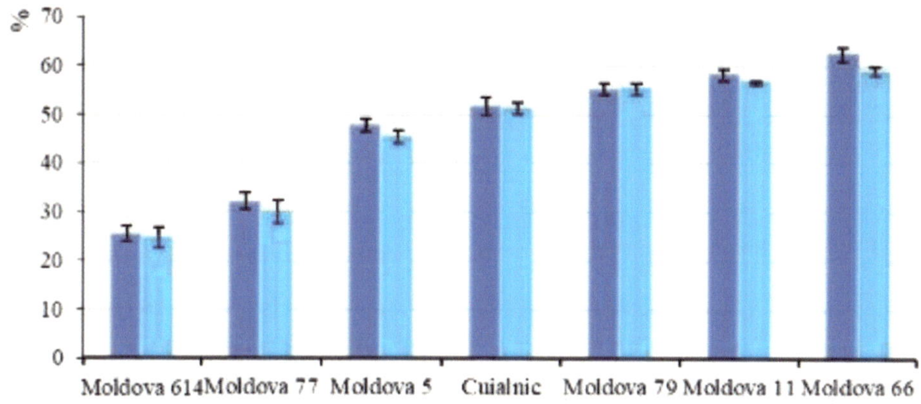

Figure 5.
The influence of the shock with temperature -7°C during 16 h on the germination percentage of the seed of different wheat varieties reproduced in 2019 (diagrams marked in dark blue—Left) and those that after collection, the seeds were stored for 1 year (diagrams colored in light blue—Right). After exposure to SNT, the seeds were incubated in the dark, temperature 24°C, for 5 days for germination. In the control variants, the germination percentage of the seeds of all studied varieties varied between 98 and 100%.

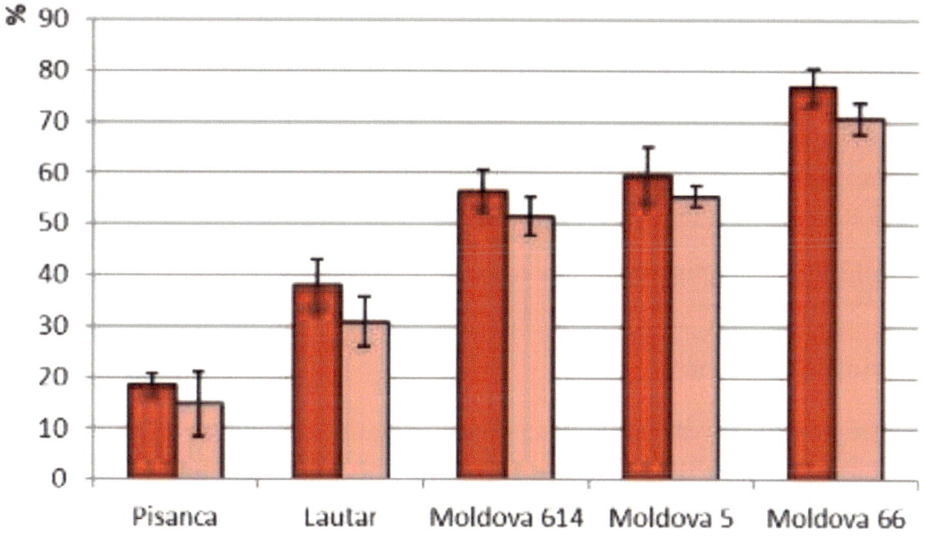

Figure 6.
The influence of the shock with temperature 50°C during 30 min on the germination percentage of the seed of different wheat varieties reproduced in 2019 (diagrams are marked in dark red—Left) and those that after collection the seeds were stored for 1 year (diagrams are marked in light red—Right). After exposure to HS, the seeds were incubated in the dark, temperature 24°C, for 5 days for germination.

(in 2020) was equal to 0.9980. At the same time, the correlation coefficient between the germination percentage of HS exposed seeds immediately after harvesting, and the value of the difference between the germination percentage of HS exposed seeds directly after harvesting and 1 year later was equal 0.2994.

In general, the above-exposed data give the possibility to suggest that the value of decreasing the percentage of seeds germination after 1 year of storage tends to be practically independent of the level of genotype primary resistance to SNT or HS at the initiation of storage. However, the significance of the small positive correlations between SNT or HS resistance, determined at the beginning of storage and the value of its decrease 1 year later, requires further research. In our view, the special attention requires additional experimental analysis of the seeds of wheat varieties with the maximum or minimum values of decreasing the primary resistance to SNT or HS after 1 year of storage. We believe that the key to understanding the significant difference between the rates of decline in these genotypes of seed resistance to HS or SNT after storage for 1 year may be associated with differences in the inheritance of acquired traits.

4. Conclusions

Exposure to HS or SNT of different wheat varieties seeds, prepared for germination in particular conditions, due to the elimination of the influence of avoidance mechanisms and adaptation processes carried out during ontogenesis, give the possibility to distribute genotypes according to their primary resistance to high temperatures frost.

Seeds of different wheat genotypes differ by their primary resistance to HS and SNT. After exposure to HS or SNT, the reduction of wheat seed germination was all the more significant, the lower the primary resistance of the genotype to the action of extreme temperatures. Accelerated HS and SNT resistance determination are practical approaches for managing existing wheat genotypes and optimizing the available germplasm screening methods by laboratory and field testing. Using the proposed guidelines in breeding and agronomy will depend on their application with other physiological and molecular methods. Thus, it is possible to determine the influence of genetic and epigenetic heredity on wheat genotypes' primary and adaptive resistance to high temperatures and frosts.

The environmental conditions of winter wheat reproduction influenced the seeds' primary resistance to HS or SNT. Seeds storage during a year, their resistance to extreme temperatures tended to decrease inversely proportional rate to that of the values of their primary resistance.

Acknowledgements

The research was carried out in the State Programs 20.80009.7007.07 by the financial support of the National Agency for Research and Development of the Republic of Moldova.

Author details

Alexandru Dascaliuc
Institute of Genetics, Physiology and Protection of Plants, Chisinau, Moldova

*Address all correspondence to: dascaliuca@yahoo.com

References

[1] National Academy of Sciences. Climate Change: Evidence and Causes. Washington, DC: The National Academies Press; 2020. pp. 1-36

[2] Akter N, Islam MR. Heat stress effects and management in wheat. A review. Agronomy for Sustainable Development. 2017;**37**:17. DOI: 10.1007/s13593-017-0443-9

[3] Dascaliuc A, Ivanova R, Arpentin G. Systemic approach in determining the role of bioactive compounds. In: Pierce GN, Mizin VI, Omelcienko A, editors. Advanced Bioactive Compounds Countering the Effects of Radiological, Chemical, and Biological Agents, Strategies to counter biological damage; Series: NATO Science for Peace and Security Series A: Chemistry and Biology. Berlin, Germany: Springer; 2013. pp. 121-131

[4] Levitt J. Chilling, Freezing, and High Temperature Stresses. 2nd ed. Cambridge, Massachusetts: Academic Press; 1980. p. 510

[5] Hassan MA, Xiang C, Farooq M, et al. Cold stress in wheat: Plant acclimation responses and management strategies. Frontiers in Plant Science. 2021;**12**:676884. DOI: 0.3389/fpls.2021.676884

[6] Skinner DZ. Genes upregulated in winter wheat (*Triticum aestivum* L.) during mild freezing and subsequent thawing suggest sequential activation of multiple response mechanisms. PLoS One. 2015;**10**(7):e0133166. DOI: 10.1371/journal.pone.0133166

[7] Frederiks TM, Christopher JT, Mark W, Sutherland MW, et al. Post-head-emergence frost in wheat and barley: Defining the problem, assessing the damage, and identifying resistance. Journal of Experimental Botany. 2015;**66**(12):1-12. DOI: 10.1093/jxb/erv088

[8] Bewley JD. Seed germination and dormancy. The Plant Cell. 1997;**9**:1055-1066. DOI: org/10.1105/tpc.9.7.1055

[9] Jaligot E, Rival A. Applying epigenetics in plant breeding: Balancing genome stability and phenotypic plasticity. In: Advances in Plant Breeding Strategies: Breeding, Biotechnology, and Molecular Tools. Vol. 1. Berlin, Germany: Springer; 2015. pp. 159-192. DOI: org/10.1007/978-3-319-22521-0_6

[10] Begcy K, Dresselhaus T. Epigenetic responses to abiotic stresses during reproductive development in cereals. Plant Reproduction. 2018;**31**:343-355. DOI: 10.1007/s00497-018-0343-4

[11] Clewer AG, Scarisbrick DH. Practical Statistics and Experimental Design for Plant Crop Science. John Wiley & Sons, Ltd: Chichester, New York; 2001. p. 332

[12] Mendanhaa T, Rosenqvist E, Hyldgaardet B, et al. Heat priming effects on anthesis heat stress in wheat cultivars (*Triticum aestivum* L.) with contrasting tolerance to heat stress. Plant Physiology and Biochemistry. 2018;**132**:213-221. DOI: org/10.1016/j.plaphy.2018.09.002

[13] Avanzy S, Brunory A, D'Aamato F. Sequential development of meristems in the embryo of *Triticum durum*. A DNA autoradiographic and cytophotometric analysis. Developmental Biology. 1969;**20**:368-377. DOI: 10.1016/0012-1606(69)90020-7

[14] Dascaliuc A, Zdioruc N, Ralea T. Determination of *Triticum aestivum* L. primary resistance to high temperature. Plant Physiology and Genetics. 2021;**53**:336-345. DOI: org/10.15407/frg2021.04.336

[15] Mahfoozi S, Limin AE, Fowler DB. Developmental regulation of

low-temperature tolerance in winter wheat. Annals of Botany. 2001;**87**:751-757. DOI: org/10.1006/anbo.2001.1403

[16] Chen D, Osborm DJ. Ribosomal genes and DNA replication in germinating wheat embryos. Nature. 1970;**225**:336-340. DOI: 10.1038/225336a0

[17] Mory YY, Chen D, Sarid S. Onset of deoxyribonucleic acid synthesis in germinating wheat embryos. Plant Physiology. 1972;**49**:20-23. DOI: 10.1104/pp.49.1.20

[18] Crowe JH, Crowe LM, Mouradian R. Stabilization of biological membranes at low water activities. Cryobiology. 1983;**20**:346-356. DOI: 10.1016/0011-2240(83)90023-8

[19] Reynolds M, Langridge P. Physiological breeding. Current Opinion in Plant Biology. 2016;**31**:162-171. DOI: 10.1016/j.pbi.2016.04.00

[20] Lopes MS, Rebetzke GJ, Reynolds M. Integration of phenotyping and genetic platforms for a better understanding of wheat performance under drought. Journal of Experimental Botany. 2014;**65**:6167-6177. DOI: 10.1093/jxb/eru384

[21] Richards RA. Physiological traits used in the breeding of new cultivars for water-scarce environments. Agricultural Water Management. 2006;**80**:197-211. DOI: 10.1016/j.agwat.2005.07.013

[22] Pingali PL. Green Revolution: Impacts, limits, and the path ahead. PNAS. 2012;**109**:12302-12308. DOI: org/10.1073/pnas.0912953109

[23] Singh A, Singh D, Kang JS, Aggarwal N. Management practices to mitigate the impact of high temperature on wheat: A review. IIOAB Journal. 2011;**2**(7):11-22

[24] Yakhin OI, Lubyanov AA, Yakhin IA, et al. Biostimulants in plant science: A global perspective. Frontiers in Plant Science. 2017;**7**(2049):32. DOI: org/10.3389/fpls.2016.02049

[25] Rouphael Y, Giuseppe C. Toward a sustainable agriculture through plant biostimulants: From experimental data to practical applications. Agronomy. 2020;**10**(1461):10. DOI: 10.3390/agronomy10101461

[26] Song J, Irwin J, Dean C. Remembering the prolonged cold of winter: Minireview. Current Biology. 2013;**23**:R807-R811. DOI: org/10.1016/j.cub.2013.07.027

[27] Arif MAR, Nagel M, Lohwasser U, Börner A. Genetic architecture of seed longevity in bread wheat (Triticum aestivum L.). Journal of Biosciences. 2017;**42**(1):81-89. DOI: 10.1007/s12038-016-9661-6

[28] Arif MAR, Nagel M, Neumann K, et al. Genetic studies of seed longevity in hexaploid wheat using segregation and association mapping approaches. Euphytica. 2012;**186**:1-13. DOI: 10.1007/s10681-011-0471-5

[29] van Treuren R, Bas N, Kodde J, et al. Rapid loss of seed viability in ex situ conserved wheat and barley at 4°C as compared to −20°C storage. Conservation Physiology. 2018;**6**(1):10. DOI: 10.1093/conphys/coy03

[30] Shvachko NA, Khlestkina EK. Molecular genetic bases of seed resistance to oxidative stress during storage. Vavilovskii Zhurnal Genetiki i Selektsii. 2020;**24**:451-458. DOI: 10.18699/VJ20.47-o

Chapter 5

Abiotic Stress Response in *Brachypodium*

Mhemmed Gandour

Abstract

Understanding the mechanisms of physiological response in plants is crucial to building sustainable agriculture, especially under the current worldwide climate and environmental crises. Thus, plants that successfully acclimate to stress can decrease growth under stressful conditions. *Brachypodium*, an undomesticated grass species with close evolutionary relationships to wheat and barley, is a promising model organism of crop research. It can grow under various conditions and possess specific adaptations or tolerance mechanisms. Hence, it promises to greatly accelerate the process of gene discovery in the grasses and to serve as bridges in the exploration of panicoid and pooid grasses, arguably two of the most important clades of plants from a food security perspective. *Brachypodium* could hence efficaciously acclimate to the drought, salinity, cold, heat, and nutrient stress variations by reversible hypo (hyper)-activation of specific genes or sustaining transcription states as well as by reducing growth and osmotic adjustment. Nonetheless, *B. stacei* and *B. hybridum* have more plasticity and more adaptiveness than *B. distachyon* to abiotic stress. This review will describe advancements in knowledge of the physiological and metabolic adjustments that are needed for abiotic stress tolerance.

Keywords: model plant, *Brachypodium*, abiotic stress, physiological response, gene expression

1. Introduction

Throughout the last fifty years, the global climate is changing at an exceptional rate. Simultaneously, the world population has known a significant increase (about twice) accompanied by a considerable increase (3 times) in cereal production, reaching 2.5 billion tn [1]. This population will certainly continue to grow to reach 9.7 billion inhabitants in 2050 [2], and the problem is that we will have to double or even triple agricultural production. This poses a serious problem for food security which, according to the FAO [3], is based not only on a sufficient supply of quantity but also, healthful and active life for humans. Consequently, ensure an increase in this production at the rate of the growth of the population remains an important challenge to which researchers must act, especially since we are aware that the known solutions to increase the productivity of agriculture in the 20th century, including the intensive use of fertilizers, are currently showing their limits.

Cereals are by far the most important food resource in the world, either for human consumption or for animal feed. At the start of the 21st century, they still provide almost half of humanity's food calories and will undoubtedly be brought to play a fundamental role in the face of the demographic and environmental

IntechOpen

challenges of the century. Average cereal yields thus fell from 1.3 to 3.5 t ha^{-1} at the global level between 1969 and 2009 [4].

Understanding the mechanisms of physiological response in plants is crucial to building sustainable agriculture, especially under the current worldwide climate and environmental crises. Thus, plants that effectively acclimate to stress can summarize growth under stressful conditions. *Brachypodium,* an undomesticated grass species with close evolutionary relationships to wheat and barley, is a promising model organism of crop research. It can grow under various conditions and possess specific adaptations or tolerance mechanisms. Hence, it promises to greatly accelerate the process of gene discovery in the grasses and to serve as bridges in the exploration of panicoid and pooid grasses, arguably two of the most important clades of plants from a food security perspective. In this review, the diverse physiologic and metabolic responses identified in *Brachypodium* so far are discussed. We also describe and discuss the current and future development of computational tools with a focus on abiotic stress-tolerance trait interactions.

2. Drought stress

Plants respond to drought stress through various crosstalk pathways. In *Brachypodium*, Verelst showed that drought stress mainly affects the final cell size, not the cell number (6% of reduction of the cell number vs. 35% of the decrease in the cell size). Thus, cell extension is affected by drought, while cell proliferation is not, which is in sharp disparity to previous annotations made in other plant species such as barley, maize, rice, wheat, and *Arabidopsis* suggesting that *Brachypodium* possesses mechanisms to defend its dividing cells against the negative impact of drought stress [5]. In addition, the natural genetic variation revealed that *Brachypodium* deals with drought stress through the combination of natural selection on standing intra-population genetic variation and phenotypic plasticity [6]. The strong natural variation in drought resistance was subsequently used to reveal physiological and metabolic mechanisms of *Brachypodium* response to drought stress. In fact, the drought responses in *Brachypodium* were characterized by changes in amino acids, boosting the glutamine that could be functioning as a stress signal. There were also variations in sugars that were appropriate to be an osmotic counter to drought, and changes in bioenergetic metabolism [7]. As well, the drought response is greatly dependent on the developmental stage. Moreover, molecular studies have identified that gene expression is modulated in the proliferation zone and is differentially expressed in the cell expansion and mature zones. In fact, the effect of severe drought on gene expression was most pronounced in the mature leaf zone, where it has been detected significant up-or down-regulation transcripts.

3. Cold

B. distachyon can serve as an attractive model for specific molecular mechanisms implicated in low-temperature responses in core Pooideae species. It encompasses cold-responsive IRIP genes which have evolved through *Brachypodium* specific gene family expansions [8]. In fact, a large cold-responsive CBF3 subfamily was identified in *B. distachyon*, whereas CBF4 homologs are absent from the genome. In addition, growing under cold conditions lead to the acquisition of novel and targeted cold-induced transcriptional responses by inducing transcriptional responses typical of cold acclimation, including the activation of the transcription factors,

C-repeat binding factors (CBFs), and structural genes IRI, COR410, and COR413 [9]. *B. distachyon* could hence efficaciously acclimate to the cold treatment and displays reversible hypoactivation of cold-regulated genes as well as it can entirely acclimate by resuming growth under diurnal-freezing conditions [9, 10]. In fact, by regulating transcriptional adaptation, transcription memory provides plasticity to *B. distachyon*'s stress responses, to develop a freezing tolerant morphology during cold acclimation.

4. Heat

The allopolyploid grass *Brachypodium hybridum* and its progenitor *Brachypodium stacei* exhibit long-term heat stress tolerance, unlike its other ancestor, *Brachypodium distachyon* [11]. Hence, these differences were explained by the fact that *B. hybridum* and *B. stacei* sustained transcriptional states under enduring stress at a similar amplitude than those under normal environments but significantly altered their transcriptome in response to heat after short-term stress whereas *B. distachyon* showed similar expression patterns between normal and heat stress conditions in both short and long-terms treatments [11]. Overall, it has been suggested that after branching out from the common ancestor and during the adaptation process, the heat acclimatization function in *B. distachyon* might have been lost. Thus, the heat-adaptive attribute in the *B. stacei* genome may perhaps influence the subsistence of both individual plants and hybrid progeny under heat stress environments.

5. Salinity

Among the various abiotic strains, salinity causes major limitations for food production, since it presents a multifold challenge to all organisms in terms of osmotic imbalance, ionic disequilibria, and generation of toxic metabolites. It limits crop yield and reduces the use of cultivated land. Plants respond to salt stress through the transcription and translation of response-associated genes, which is a complex mechanism that implicates various crosstalk pathways. In addition, post-translational phosphorylation modification can control protein functions to respond to abiotic stress [12]. In *Brachypodium* and at the protein expression level, most of the differentially expressed proteins (DEPs) were down-regulated under stress conditions. Nonetheless, it principally acted as functional proteins, however, most of the phosphoproteins were categorized as regulatory proteins, suggesting that *Brachypodium* can react and defend against salt strain by two methods: (1) through phosphorylation variation changes, mainly involving signal transduction, transcription/translation, and transport; and (2) via protein expression changes, which mainly happen in photosynthesis and energy production [13]. Furthermore, 101 NAC genes have been identified in *B. distachyon,* among which BdNAC003 and BdNAC044 are stimulated by high salt stress [14]. Wang et al. [15] have characterized 44 BdSnRKs in *B. distachyon*, and the overexpression of BdSnRK2.9 in tobacco enhanced its tolerance to drought and salt stresses. At the expressions of transcription factor (TF) level, family members such as MYB, bHLH, and AP2/ERF were increased under salt stress, regulating the response of *Brachypodium* to salt stress. In addition, under 200 mM NaCl stress, the soluble sugar and proline content of *Brachypodium distachyon* increased significantly [16] suggesting that the osmotic adjustment is an imperative mechanism to avoid salt stress. Consequently, it completes osmotic adjustment to reduce salt damage by selectively captivating inorganic ions and accumulating organic solutes that are nontoxic to cells. One of the physiological responses regulated by ABA is associated with stomatal closure, which can avoid excessive transpiration and reduce water loss [17]. In fact, preserving water stability in plant cells is a vital strategy for plants to shield against salt stress, just as most halophytes have the characteristics of succulents [18]. For non-succulent glycophytes, it is also an imperative method to alleviate plant ion toxicity and osmotic shock by controlling the stomatal opening and wax metabolism of epidermal cells in plant leaves to reduce water loss [15, 19].

6. Nutrient availability

Nutrient stress (deficiency or excess) seriously affects plant growth, yield, and quality. *Brachypodium distachyon* (*Brachypodium*) has been proposed as a good model to enhance this knowledge in C3 temperate cereals [20]. Thus, it interacts with increased nutrient concentration by increasing biomass [21]. In fact, P and N supply had great effects on the root system of *B. distachyon*. The most noticeable effect of both N and P scarcity on *B. distachyon* was that only Leaf Node Root was significantly reduced by minor nutrient supply [21] which suggests that *B. distachyon* cannot be considered "low-P-N adapted". Comparable observations have been made in wheat in which seminal roots were much less sensitive to N, P, and K deficiencies than nodal root growth and emergence [22]. By comparing plant growth at diverse concentrations of P and N, it appeared that *Brachypodium* required approximately three to four times more N than P for the same biomass production. Overall, *Brachypodium* showed plasticity in its biomass allocation pattern in response to variable P and N conditions, specifically by prioritizing root expansion overshoot productivity under poorly soluble P or N conditions (shoot productivity was depressed in *Brachypodium distachyon* while the root system development was sustained) [23]. *B. distachyon* was revealed as a good model to study ammonium nutrition since it responded likewise to other monocot crops, but with less complexity. The plants increased the storage of $NH4^+$ in roots, as well as the synthesis of amino acids and proteins. Indeed, it seemed to be moderately tolerant to ammonium. Notably, 1 mM was considered an N-sufficient condition, since expanding NO_3^- supply to 2.5 mM did not further increase plant biomass. Nevertheless, when NH_4^+ was elevated to 2.5 mM plants showed moderate indicators of ammonium toxicity in terms of leisurely growth. In addition, the root system is shown as a physiological barrier acting as a reservoir for free NH_4^+ and increasing NH_4^+ assimilation to amides.

At the molecular level, an extensive BLAST search was carried to identify putative orthologues of the *Arabidopsis NRT2* genes in the wholly sequenced *Brachypodium* genome. Seven genes encoding putative high-affinity nitrate transporters (*BdNRT2*) were identified. Only *BdNRT2.1* and *BdNRT2.2* were highly expressed in the root and classified as inducible genes, suggesting they are likely the main contributors to root nitrate uptake. *BdNRT2.5* has shown to be stifled by nitrate resupply however further members were constitutively expressed in the root. Conspicuously, great ammonium concentrations also induced analogous gene expression regulation, suggesting *BdNRT2* gene expression was also governed by inside nitrogen status, not just outside nitrate concentrations [24]. Additionally, *BdNRT2.1* was also strongly expressed in the stem, indicating that it has useful roles other than nitrate uptake.

Concerning zinc, *Brachypodium* exhibited the typical performance of a zinc-sensitive, excluder plant [25]. It prioritized shoot zinc accumulation upon deficiency and majorly retaining zinc in roots upon excess, in both cases to preserve the photosynthetic function in leaves. In addition, clear repression of vegetative growth was accompanied by increased leaf number suggesting that in order to optimize nutrient use efficiency in shoot and maintain photosynthesis, plants have adjusted leaf area partitioning [26]. Deficiency and excess treatments increased lateral root number and length relative to the primary root, and nodal roots, post-embryonic shoot-born roots emerging from successive shoot nodes and a unique feature of monocots, were strongly affected. At the molecular level, The *Brachypodium* homolog of *AtbZIP19 (in Arabidopsis)*, Bradi1g30140 was previously suggested to be involved in a zinc deficiency-induced oxidative stress response [27, 28]. However, it was slightly more expressed in zinc-deficient shoots compared to control plants and displayed a very flattened V-shape dynamics upon zinc resupply. AtbZIP19 and AtbZIP23 are proposed to be specialized in either roots or shoots, respectively [29, 30]. In *Brachypodium*, *BdbZIP9* was more expressed in shoots than roots. Interestingly, another *bZIP* gene, Bradi1g29920, was majorly expressed in roots. Moreover, 113 TFs from diverse families such as bZIP (9 genes), bHLH (11 genes), MYB (22 genes), AP2 (24 genes), and WRKY (25 genes), were known DEG. But no one of them are homologs of identified zinc regulatory genes and formed potential candidates for a role in zinc homeostasis regulation in grasses [31].

7. Conclusions and future perspectives

Engineered plants will cover the way for future strategies to adapt them for higher biomass production to meet the demands of a growing population in a changing climate scenario. Similarly, understanding the mechanisms of physiological and molecular response in plants will uncover the complexity of the dynamic changes during cell wall development and abiotic stress response. In parallel, phenomics can help in identifying the key factors affecting plant growth and health, and subsequently plant productivity, since this technology allows the non-destructive screening of hundreds of plants in a very short time. Thus, employing the developing omics approaches especially the signaling cascades in response to abiotic stresses in tolerant plants will help to manipulate susceptible crop plants and increase agricultural productivity in the near future. Moreover, GWAS will contribute to better understanding the abiotic stress response.

Author details

Mhemmed Gandour[1,2]

1 Faculty of Sciences and Technologies of Sidi Bouzid, Kairouan University, Tunisia

2 Center of Biotechnology of Borj-Cedria, Tunisia

*Address all correspondence to: gandourmed@yahoo.fr

References

[1] FAO. The State of Food and Agriculture. Climate change, agriculture and food security, Italy; 2016. Available from: https://www.fao.org/publications/sofa/sofa2016

[2] United Nations New York, Department of Economic and Social Affairs. World Population Prospects 2019; 2019. Available from: https://population.un.org/wpp/Publications/Files/WPP2019_Highlights.pdf

[3] FAO. L'état de l'insécurité alimentaire dans le monde, Rome; 2009. Available from: https://www.fao.org/3/i0876f/i0876f.pdf

[4] FAO. 2011. FAOSTAT Online Database. Available from: http://faostat.fao.org/. [Accessed December 2011]

[5] Verelst B. Managing inequality: The political ecology of a small-scale fishery, Mweru Luapula, Zambia. Journal of Political Ecology. 2013;**20**:14-36

[6] Ibrahim Y, Neji M, Taamalli W, Abdelly C, Gandour M. The genetic variation in response to drought in Tunisian populations of *Brachypodium hybridum* (Poaceae): An interplay between natural selection and phenotypic plasticity. Environmental and Experimental Botany. 2020;**179**:104234

[7] Skalska A, Beckmann M, Corke F, Tuna GS, Tuna M, Doonan JH, et al. Metabolomic variation aligns with two geographically distinct subpopulations of *Brachypodium Distachyon* before and after drought stress. Cell. 2021;**10**:683

[8] Li C, Rudi H, Stockinger EJ, Cheng H, Cao M, Fox SE, et al. Comparative analyses reveal potential uses of *Brachypodium distachyon* as a model for cold stress responses in temperate grasses. BMC Plant Biology. 2012;**12**:1-15

[9] Mayer BF, Charron JB. Transcriptional memories mediate the plasticity of cold stress responses to enable morphological acclimation in *Brachypodium distachyon*. New Phytologist. 2021;**229**:1615-1634

[10] Crisp PA, Ganguly D, Eichten SR, Borevitz JO, Pogson BJ. Reconsidering plant memory: Intersections between stress recovery, RNA turnover, and epigenetics. Science Advances. 2016; **2**:e1501340

[11] Takahagi K, Inoue K, Shimizu M, Uehara-Yamaguchi Y, Onda Y, Mochida K. Supporting Data for 'Homoeolog-Specific Activation of Genes for Heat Acclimation in the Allopolyploid Grass *Brachypodium hybridum*'. GigaScience Database. 2018. DOI: 10.5524/100413

[12] Barjaktarović Ž, Schütz W, Madlung J, Fladerer C, Nordheim A, Hampp R. Changes in the effective gravitational field strength affect the state of phosphorylation of stress-related proteins in callus cultures of *Arabidopsis thaliana*. Journal of Experimental Botany. 2009;**60**:779-789

[13] Lv B, Nitcher R, Han X, Wang S, Ni F, Li K, et al. Characterization of *FLOWERING LOCUS T1* (*FT1*) gene in *Brachypodium* and wheat. PLoS One. 2014;**9**:e94171

[14] You J, Zhang L, Song B, Qi X, Chan Z. Systematic analysis and identification of stress-responsive genes of the NAC gene family in *Brachypodium distachyon*. PLoS One. 2015;**10**:0122027

[15] Wang L, Hu W, Sun J, Liang X, Yang X, Wei S, et al. Genome-wide analysis of SnRK gene family in *Brachypodium distachyon* and functional characterization of BdSnRK2.9. Plant Science. 2015;**237**:33-45

[16] Guo X, Wang Q, Liu Y, Zhang X, Zhang L, Fan S. Screening of salt stress responsive genes in Brachypodium

distachyon (L.) Beauv. by transcriptome analysis. Plants. 2020;**9**:1622

[17] WasKiewicz A, Beszterda M, Golinski P. In: Ahmad P, Azooz MM, Prasad MNV, editors. Salt Stress in Plants. New York: Springer; 2013

[18] Zhang X, Yao Y, Li X, Zhang L, Fan S. Transcriptomic analysis identifies novel genes and pathways for salt stress responses in *Suaeda salsa* leaves. Scientific Reports. 2020;**10**:4236

[19] Ma J, Gao X, Liu Q, Yun S, Zhang D, Jiang L, et al. Overexpression of TaWRKY146 increases drought tolerance through inducing stomatal closure in *Arabidopsis thaliana*. Frontiers in Plant Science. 2017;**8**:2036

[20] Girin T, David LC, Chardin C, Sibout R, Krapp A, Méry SF, et al. *Brachypodium*: A promising hub between model species and cereals. Journal of Experimental Botany. 2014;**65**:5683-5696

[21] Poiré R, Chochois V, Sirault XRR, Vogel JP, Watt M, Furbank RT. Digital imaging approaches for phenotyping whole plant nitrogen and phosphorus response in *Brachypodium distachyon*. Journal of Integrative Plant Biology. 2014;**56**:781-796

[22] Tennant D. Root growth of wheat. 1. Early patterns of multiplication and extension of wheat roots including effects of levels of nitrogen, phosphorus and potassium. Australian Journal of Agricultural Research. 1976;**27**:183-196

[23] Baudson C, Delory BM, Spaepen S, Jardin PD, Delaplace P. Developmental plasticity of *Brachypodium distachyon* in response to P deficiency: Modulation by inoculation with phosphate-solubilizing bacteria. Plant Direct. 2021:1-17

[24] Wang J, Hüner N, Tian L. Identification and molecular characterization of the *Brachypodium distachyon* NRT2 family, with a major role of BdNRT2.1. Physiologia Plantarum. 2019;**165**:498-510

[25] Krämer U. Metal hyperaccumulation in plants. Annual Review of Plant Biology. 2010;**61**:517-534

[26] Smith DD, Sperry JS, Adler FR. Convergence in leaf size versus twig leaf area scaling: do plants optimize leaf area partitioning? Annals of Botany. 2017;**119**:447-456

[27] Glover-Cutter KM, Alderman S, Dombrowski JE, Martin RC. Enhanced oxidative stress resistance through activation of a zinc deficiency transcription factor in *Brachypodium distachyon*. Plant Physiology. 2014; **166**:1492-1505

[28] Martin RC, Vining K, Dombrowski JE. Genome-wide (ChIP-seq) identification of target genes regulated by BdbZIP10 during paraquat-induced oxidative stress. BMC Plant Biology. 2018;**18**

[29] Sinclair SA, Senger T, Talke IN, Cobbett CS, Haydon MJ, Kraemer U. Systemic upregulation of MTP2- and HMA2-mediated Zn partitioning to the shoot supplements local Zn deficiency responses of Arabidopsis. The Plant Cell. 2018;**30**:2463-2479

[30] Arsova B, Amini S, Scheepers M, Baiwir D, Mazzucchelli G, Carnol M, et al. Resolution of the Proteome, Transcript and Ionome Dynamics Upon Zn Re-supply in Zn-Deficient Arabidopsis. bioRxiv; 2019. p. 600569

[31] Amini S, Arsova B, Gobert S, Carnol M, Bosman B, Motte P, et al. Transcriptional Regulation of *ZIP* Genes is Independent of Local Zinc Status in *Brachypodium* Shoots Upon Zinc Deficiency and Resupply. 2021;**44**:3376-3397

Chapter 6

Mechanistic Role of Reactive Oxygen Species and Its Regulation *via* the Antioxidant System under Environmental Stress

Ambreen Bano, Anmol Gupta, Smita Rai, Touseef Fatima, Swati Sharma and Neelam Pathak

Abstract

The antioxidant potential is referred to as compounds that are capable of protecting the biological system against the deleterious effect of reactions involving reactive oxygen species (ROS). ROS are toxic byproducts of oxygen metabolism. ROS have a significant role in plant signaling, growth, development, and majorly in response to environmental fluctuations. The ROS family plays a double role under various environmental stress conditions. In various key physiological phenomena, they act as secondary messengers and induce oxidative damage. ROS led to cellular damages that manifest themselves in degradation of biomolecules, which eventually amalgamate to cellular death in plants. To assure survival, plants have developed efficient antioxidant machinery having two branches, that is, an enzymatic and a nonenzymatic antioxidant. This chapter will emphasize the various types of ROS, their sites of cellular production, targets, and scavenging mechanisms mediated by antioxidants in abiotic stress. Such profound knowledge will let us build strategies against environmental stress.

Keywords: ROS, abiotic stress, antioxidant, free radical

1. Introduction

Plants are continually susceptible to environmental changes, prompting them to regulate their metabolism in such a way as to maintain a constant balance between the generation of energy and its consumption. This delicate balance is majorly dependent on a network signaling that mainly coordinates among the key operations in plant life including dark respiration, photorespiration, and photosynthesis, all of these activities are linked by reductants, substrate, energy, and electron transfer [1, 2]. Plant organelle metabolic pathways are sensitive to climate change and metabolic inequities in cells that can cause oxidative stress by boosting the oxidation of cellular components, production as well as accumulation of reactive oxygen species (ROS), impeding metabolic activities, and affecting organelle integrity [3, 4].

Environmental abiotic stresses, such as chilling, salinity, harsh temperature, drought, toxic metals or metalloids, flooding/waterlogging, and ultraviolet (UV)

radiation, have become more common as a result of abrupt and extreme climate change [5]. The escalation of various abiotic stresses emerged as a key threat to agricultural crop production. Furthermore, an excess of ROS such as free radical alkoxy radical ($RO^{\bullet}$); hydroperoxyl radical ($HO_2^{\bullet}$); hydroxyl radical ($OH^{\bullet}$); superoxide anion ($O_2^{\bullet}$); and nonradical molecules (singlet oxygen, $^1O_2$, and hydrogen peroxide, H_2O_2) causes plant oxidative stress [6]. The principal sites of cellular ROS formation are mitochondria, chloroplast, apoplast, plasma membranes, and peroxisome [7]. Although as a component of regular cellular metabolism ROS are produced in plants, their overabundance due to several stresses severely destroys essential cellular constituents such as DNA, proteins, carbohydrates, lipids, and so on due to their highly reactive nature [8]. Nonenzymatic, low-molecular substances such as ascorbic acid (AA), reduced glutathione (GSH), carotenoids, tocopherol, proline, phenols, and flavonoids as well as guaiacol peroxidase (GPX), peroxidase (APX), superoxide dismutase (SOD), catalase (CAT), and glutathione-S-transferase (GST). The pervasiveness of both of the antioxidant machinery's arms supports the necessity for ROS detoxification for cellular viability [9].

1.1 Types of ROS

ROS are generally a molecular O_2 that has been incompletely activated or reduced or the principal product or the O_2-containing molecule byproduct that has an elevated reactivity than ambient O_2. ROS are produced by the electrons from the O_2 molecule or transfer of energy. The most prevalent cellular ROS in plants are $OH^{\bullet}$, $O_2^{\bullet}$, $^1O_2$, and H_2O_2. Cells generate both nonradical and free radical ROS. Free radicals include $OH^{\bullet}$, $O_2^{\bullet}$, $ROO^{\bullet}$, and $RO^{\bullet}$, while the nonradicals include $^1O_2$ and H_2O_2. The other ROS nonradicals found in plants include excited carbonyl (RO^{*}) and hypochlorous acid (HOCl). Furthermore, few acidic molecules such as hypoiodous acid (HOI), hypobromous acid (HOBr), and hypochlorous acid (HOCl) and radical compound ($CO_3^{\bullet}$) are incorporated into ROS. In addition, biological systems may contain alkoxyl radical ($LO^{\bullet}$), peroxyl radical ($LOO^{\bullet}$), hydroperoxyl radical ($HO_2^{\bullet}$), peroxynitrite (HNO_3), trichloromethyl peroxyl radical ($Cl_3COO^{\bullet}$), and ozone (O_3).

1.1.1 Superoxide radical ($O_2^{\bullet-}$)

In the chloroplasts, ROS are constantly produced as a result of partial O_2 reduction or energy transfer to O_2. During the noncyclic electron transport chain (ETC), $O_2^{\bullet-}$ is generated mostly in the thylakoid localized PSI and other cellular compartments. When cytochrome c oxidase reacts with O_2, H_2O is normally produced. Occasionally, O_2 reacts with the various ETC components to produce $O_2^{\bullet}$. It is typically the first ROS to develop. $O_2^{\bullet-}$ can potentially undergo a further reaction, resulting in the formation of other ROS [10].

1.1.2 Singlet oxygen ($^1O_2$)

It is an unusual ROS that is produced *via* the chlorophyll reaction (in the antenna system, in the triplet state) rather than electron transport to O_2. Heavy metals, salinity, and drought all cause stomatal closure, resulting in lack of intracellular CO_2 concentration. Facilitating the production of $^1O_2$ and causing significant damage to both the photosystems (PSI and PSII) put the whole photosynthetic apparatus in danger, causing damage to a wide spectrum of the target. These substances include pigments, proteins, lipids, and nucleic acids and are the primary ROS responsible for light-induced loss of PSII function, resulting in cellular death [11].

DOI: http://dx.doi.org/10.5772/intechopen.101045

1.1.3 Hydrogen peroxide (H_2O_2)

It is created in plant cells under normal conditions as well as in response to oxidative stresses (drought, cold, UV radiation, bright light, pathogen infection, and wounding). $O_2^{\bullet-}$ undergoes both univalent protonation and reduction, and a moderately reactive H_2O_2 is produced. The ER, mitochondria, ETC in the chloroplast, oxidation of fatty acids, photorespiration, and cell membrane are the key sources of H_2O_2 generation in plant cells. In plants, H_2O_2 is useful at low quantities but harmful at greater amounts. It operates as a regulatory signal for critical physiological processes such as senescence, stomatal movement, photosynthesis and photorespiration, growth as well as development at low intracellular concentrations [12].

1.1.4 Hydroxyl radical ($OH^{\bullet}$)

$OH^{\bullet}$ is the most reactive and destructive ROS. At neutral pH, it is produced *via* the Fenton reaction between H_2O_2 and $O_2^{\bullet-}$ catalyzed by transition metals such as Fe (Fe^{2+}, Fe^{3+}). $H_2O_2 + O_2^{\bullet-} \rightarrow OH^- + O_2 + OH^{\bullet}$. It is capable of causing harm to several biological components through lipid peroxidation (LPO), protein degradation, and membrane disintegration. Because no enzymatic system exists to scavenge this deadly radical, excessive $OH^{\bullet}$ buildup causes cellular death [13].

1.1.5 Peroxyl radical ($ROO^{\bullet}$)

The key chain-propagating step in lipid peroxidation and non-lipid systems is the formation of $RO_2^{\bullet}$ and $RO^{\bullet}$ radicals that can be generated by the decomposition of protein and lipid peroxides when heated or by the addition of transition metal ions. It is easy to produce peroxyl radicals by combining O_2 with carbon-centered radical $>C^{\bullet} + O_2 > C^-OO^{\bullet}$. The peroxyl radical plays an important role in the oxidation of lipids, DNA damage, changes in the protein backbone, and the degradation of food.

1.1.6 Alkoxy radical ($RO^{\bullet}$)

When lipids are oxidatively degraded or peroxidized without the help of enzymes, alkoxyl radicals are generated by the Fenton reaction, by electron reductions, or by combining two peroxyl radicals. Apoptosis and DNA alterations may result from alkyl radical oxidation. DNA damage and apoptosis can be caused by alkoxyl radicals, which are very oxidizing.

2. Sites of ROS production/generation in plant cells

It is proven that ROS can be produced in multiple places in the mitochondria, chloroplasts, plasma membranes, peroxisomes, endoplasmic reticulum, and cell wall both under normal and stressful conditions. ROS are produced primarily by peroxisomes and chloroplasts when light is present, while the mitochondria generate ROS when light is not present.

2.1 Chloroplast

Chlorophyll (chl) and light interact with each other to produce ROS in the chloroplast, which is the most important site where ROS are produced. In this case, ROS are principally generated by triplet chl and electron transport chains (ETC), including PSI and PSII. SOD converts $O_2^{\bullet-}$ into H_2O_2 under PSI in the Mehler reaction [14]. The $O_2^{\bullet-}$ and H_2O_2 generate more highly reactive $HO^{\bullet}$ when combined with metal ions such as Fe^{2+}. Various ecological stressors cause stomatal closure, resulting in a decrease in CO_2 levels, thereby causing the generation of chloroplastic ROS [3, 15].

2.2 Mitochondria

Although on a smaller scale, mitochondria are also responsible for damaging ROS production, including H_2O_2 and $O_2^{\bullet-}$. This is due to the mitochondrial ETC (mt ETC) containing enough energetic electrons to reduce O_2 and produce ROS. The two main mt ETC components responsible for the production of ROS are complexes I and III [16]. Additionally, the mitochondrial matrix contains numerous enzymes that produce ROS. Mn-SOD and APX reduce O_2 into H_2O_2 under the influence of oxygen, despite its abundance in the mitochondria. When mitochondria are under abiotic stress, ROS production increases dramatically [17].

2.3 Apoplast

The diffusible region around the plant cell membrane appears to be in charge of converting incoming CO_2 into a soluble, diffusible form that can be transported into the cytoplasm to carry out photosynthesis. During the harsh environmental situation, stress signals combined with abscisic acid make the apoplast a significant location for H_2O_2 production. Several other substances, such as pH-dependent peroxidases (POXs), polyamine oxidases, and cell-wall-linked oxidases, also generate ROS [18].

2.4 Plasma membranes

Plant cells are surrounded by a plasma membrane, which is constantly interacting with changing external conditions, thereby providing essential information for their survival.

During electrons transferring from cytosolic NADPH to O_2, either e^- dismutates spontaneously to H_2O_2 or is catalyzed by NADPH oxidase; SOD forms $O^{\bullet}_2$. The importance of NADPH oxidase in plant defense against pathogenic infection and abiotic stress conditions has been well established [19].

2.5 Cell walls

The cell wall-localized lipoxygenase (LOX) creates polyunsaturated fatty acid (PUFA) hydroperoxidation, making it an active generator of ROS such as $O_2^{\bullet}$, $OH^{\bullet}$, $^1O_2$, and H_2O_2. Using diamines or polyamines, cell wall-localized diamine oxidases produce ROS in the cell wall. During the pathogenic attack, lignin precursors undergo considerable cross-linking *via* H_2O_2-mediated pathways, resulting in the formation of recombinant lignin [20].

2.6 Endoplasmic reticulum (ER)

Cyt P450, which is located in the ER, is used to create $O_2^{\bullet-}$ *via* NADPH-mediated electron transport. RH (an organic substrate) interacts with CytP450, which is formerly reduced by a flavoprotein to form a free-radical intermediate (Cyt P450 R). This intermediate forms an oxygenated complex when it reacts with triplet oxygen ($^3O_2$) (Cyt P450-ROO). When the complex decomposes to Cyt P450-Rh, $O_2^{\bullet}$ is produced as a byproduct [21].

3. Role of ROS as messengers

ROS have been identified as the second messenger in intracellular signaling cascades that mediate a variety of plant responses, including programmed cell death, stomatal closure, gravitropism, and abiotic and biotic stress-tolerance acquisition. ROS can also affect the activity of numerous signaling components, including protein phosphatases, transcription factors, and protein kinases as well as communicate with other signal molecules and the pathway that administers response downstream of ROS. The balance between oxidant formation and antioxidant removal determines the life span, strength, and size of the ROS signaling pool [19, 20].

4. ROS and oxidative damage to biomolecules

The creation and removal of reactive oxygen species must be appropriately regulated to avoid oxidative stress. When the number of reactive oxygen species exceeds the cell's defensive systems, the cell is said to be in "oxidative stress." The balance between ROS formation and scavenging is upset in several stressful circumstances, such as salt, thirst, metal toxicity, intense light, viruses, and so on. High quantities of ROS can harm biomolecules, such as lipids, proteins, and DNA. These processes can alter inherent membrane properties including fluidity, enzyme activity loss, ion transport, protein synthesis suppression, protein cross-linking, DNA damage, etc., resulting in cell death. Redox homeostasis develops in plant cells as a result of equilibrium between the creations of ROS and the functioning of antioxidant enzymes, where a well-functioning defense system in plants maintains the right balance between ROS generation and its removal.

For proper redox signaling in the cell, it is, therefore, necessary to maintain an amount of ROS above or below the cytotoxic concentration, which is achieved by maintaining the equilibrium between ROS production and scavenging. As a result, a constant equilibrium between ROS formation and scavenging systems is maintained by cooperating with cellular redox-sensitive components to precisely adapt the downstream signaling procedures in a context-specific and cell-specific manner. Under varied abiotic stress conditions, any disruption in the equilibrium between ROS formation and scavenging by antioxidants leads to ROS excess buildup, culminating in oxidative stress [9]. Oxidative stress results in damaging the nucleic acid and protein and lipid peroxidation, thereby altering the carbohydrate metabolism and thus leading to cellular death and its dysfunction.

4.1 Lipids

Increasing ROS levels trigger lipid peroxidation in membranes of cells and organelles, affecting normal cellular activity. Oxidative stress is exacerbated by lipid peroxidation by producing radicals from lipids, which affect proteins and DNA. In stressed cells, lipid peroxidation can be used as a biomarker of ROS-mediated membrane damage. Environmental challenges have been shown to cause increased lipid peroxidation and its degradation in several plants [22].

4.2 Proteins

Besides direct changes, ROS can indirectly affect proteins. Direct modification is the process of modifying a protein's function by disulfide bond formation, nitrosylation, carbonylation, and glutathionylation. Through the breakdown of fatty acid peroxidation products, proteins are indirectly altered. The increased ROS production alters amino acid sites, fragments peptide chains, aggregates cross-linked reactions, changes the charge, and increases the proteolytic activity of proteins. The amount of carbonylated proteins in oxidatively damaged tissues is higher, a measurement of protein oxidation. Plants are reported to modify proteins in response to diverse stressors. There is a wide range of ROS attacks that can be performed on amino acids in peptides [23, 24].

4.3 Damage to nucleotides and DNA

The hydroxyl radical ($OH^{\bullet}$) is one of the damaging factors for polynucleic acids, because it changes the pyrimidine and purine structure by liberating H^+ from the C–H bonds of methyl and 2-deoxyribose group, generating deoxyribose radical, thymine glycol, hydroxyl methyl urea, and other compounds, thereby breaking double-stranded DNA into single-stranded DNA. DNA damage is caused by oxidative stress. By oxidizing deoxyribose sugar, changing nucleotide bases, cross-linking DNA, proteins, and abstracting nucleotides, ROS cause DNA nucleotide damage. Plant growth, as well as development, is influenced by DNA damage through a variety of physiological mechanisms, including abnormal protein synthesis and damage to photosynthetic proteins, among others. It also can prevent replication mistakes, signal transduction, transcription, and overall genomic instability. In addition, DNA bases are damaged and irreversible not only by direct oxidation but also by reactive intermediates (associated with ROS attack) reacting with macromolecules [25, 26].

5. Oxidative stress under abiotic stress

There are different types of abiotic stresses as depicted in **Figure 1**.

5.1 Oxidative stress under salinity

Among all above-mentioned abiotic stresses, salinity or salt stress is regarded as one of the most damaging, reducing land area and agricultural productivity. Soil salinity is a global issue that affects around 20% of irrigated land and severely lowers agricultural production [27]. Salinity has a negative impact on crop germination, yield, and vigor. Osmotic stress, nutritional ionic specificity, hormonal problems, altered physiological and metabolic processes, and, finally, oxidative damage occur when plants are exposed to high salt concentration. Some of the most typical effects of salt stress in plants are photosynthetic reduction, nutritional unavailability, cellular membrane disruption, the creation of several toxic metabolites, and eventual plant death.

Salinity stress causes overproduction of ROS that affects the plants leading to ion toxicity, nutritional inadequacy, osmotic stress, and genotoxicity that causes oxidative stress [28, 29].

5.2 Drought

Drought causes stomatal closure and decreases CO_2 fixation in plant leaves. As a result, this stress causes an imbalance between light capture and utilization, lowering

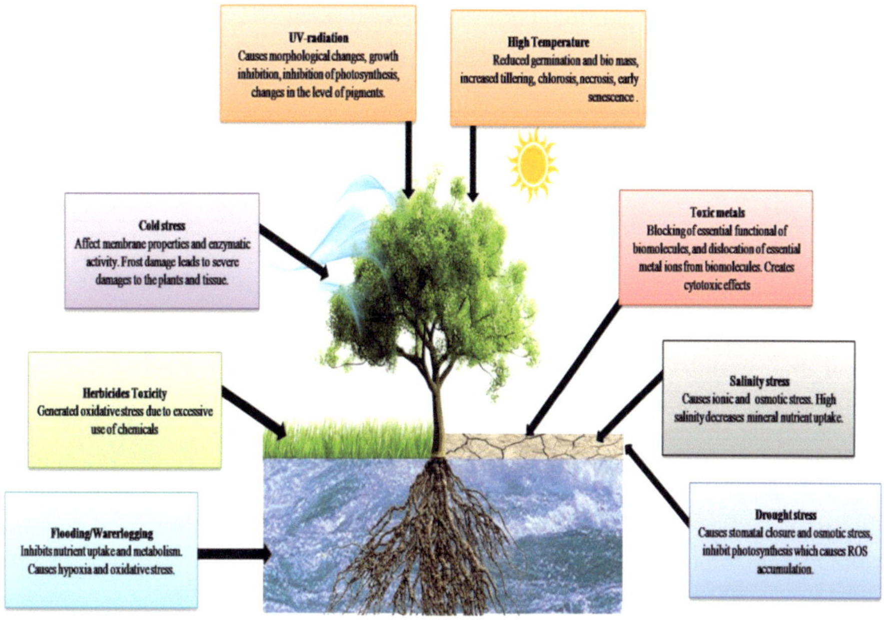

Figure 1.
Different types of abiotic stresses for plants.

the photosynthetic rate [30]. During drought, the photochemistry of chloroplasts changes, and an imbalance between electron acceptance and release results in an increase in the formation of ROS from surplus light energy in the photosystems. In reality, ROS are produced by absorbed light energy, which cannot be converted to CO_2 fixation. Photorespiration directly causes drought-induced H_2O_2 production. Furthermore, under drought stress, if chloroplasts are exposed to excessive light energy, ferredoxin levels plummet; $NADP^+$ regeneration is hampered, interfering with electron uptake, resulting in lower ETC and increased electron leakage, both of which contribute to ROS overproduction. Drought-induced LPO, malondialdehyde (MDA), and H_2O_2 accumulation cause malfunction of several physiological and cellular processes including membrane functions, water-use efficiency, stomatal conductance, carboxylation efficiency, transpiration, respiration, and photosynthesis. Drought stress increases MDA and H_2O_2 levels in several plant species, including maize, chili, rapeseed, alfalfa, soybean, and others, which, along with other harmful ROS, cause oxidative damage [31].

5.3 Metal toxicity

Toxicity from metals or metalloids not only disrupts morphophysiological features but also causes increased oxidative stress due to an imbalance between the antioxidant defense system and ROS generation. In a study, it was found that nickel (100 M) stressed *Pisum sativum* L. seedlings accumulated more LPO and H_2O_2. Such increases in oxidative stress indicators were even greater in the same crop with the same Cd content. Meanwhile, cadmium stress increases the H_2O_2 and MDA production in numerous crops. Aside from H_2O_2 and MDA, the rate of $O_2^{\bullet}$ generation in *Vigna radiata* L. was also higher [5, 32].

5.4 Flooding or waterlogging

In the natural environment, unexpected extreme climate change occurs, such as flooding, which can disrupt the natural distribution of plants and even lead to extinction. Flooding or waterlogging (WL) can cause anoxia as well as hypoxia by impeding respiration and producing toxic chemicals that impede metabolic processes. The main reasons for yield reduction are decreased growth and production of biomass, disruption in light interception, limitation of stomatal conductance and CO_2 assimilation, reduced respiration and photosynthesis, and altered accumulation of secondary metabolites. Under WL conditions, such metabolic process deficiencies result in the production of ROS and cause oxidative damage to the cell [33].

5.5 Chilling

Temperatures (1–10°C) in the chilling-sensitive plant's cells cause multiple physiological abnormalities, resulting in chilling injury and mortality in many tropical and subtropical plant species, including many vegetable species. According to a recent study, exposing chilling-sensitive plants to low temperatures disrupts all physiological functions, including water regime, mineral feeding, photosynthesis, respiration, and metabolism [34]. The reported inactivation of metabolism during chilling of chilling-sensitive plants is a complicated function of both temperature and exposure period. Plants' responses to low temperature are connected with changes in the rate of gene transcription of a variety of low-molecular-weight proteins.

Chilling stress denotes low-temperature harm without the development of ice crystals, whereas freezing stress denotes injury caused by ice formation within plant tissues. Plant species are dependable in their tolerance to chilling or freezing conditions. Chilling temperature decreases crop productivity by interfering with various components of plant growth and development. At the vegetative stage, freezing temperatures restrict seedling growth and create aberrant phenotypes, such as stem discoloration, leaf yellowing or whitening, white patches/spots, wilting, and diminished tillering. Mung bean seedling development and dry weight reduced when exposed to a chilling temperature of 6°C. These seedlings also have aberrant phenotypic characteristics. Cold (11°C) damage symptoms were detected in rice [35, 36]. In chill-affected rice plants, stunted development, leaf chlorosis, an uneven number of tillers, and deformed and discolored grain symptoms were common. Lower temperatures affected soybean growth time, biomass accumulation, harvested index, seed number, and seed weight [37].

5.6 Extremely high temperature

Rising global temperatures over the previous few decades have resulted in major crop losses in a variety of regions throughout the world. By the year 2100, global temperature is anticipated to rise by up to 2.5–5.4°C. Temperature intensity, duration, and rate of increase are all critical factors in causing plant damage. Some frequent damaging consequences of high temperature include reduced germination and biomass, increased tillering, chlorosis, necrosis, early senescence of floral buds, premature mortality, and fruit senescence. Heat stress during seed development causes structural disintegration and physiological problems, reducing germination and vigor, emergence, and seedling establishment even further. High temperatures hampered rice seed germination by lowering the levels of a collection of proteins involved in methionine metabolism, amino acid biosynthesis, energy metabolism, reserve degradation, and protein folding [38, 39].

5.7 Ultraviolet radiation

UV is a type of radiation that is produced by the sun. Solar radiation provides essential energy for plant growth and development through photosynthesis, but high light and, in particular, its integral ultraviolet (UV) fraction cause stress, potentially resulting in severe injury to plant cellular components such as DNA and protein.

The regular discharge of chlorofluorocarbons and other pollutants as a result of human activity increases the amount of UV radiation on the earth's surface, which is the primary cause of stratospheric ozone layer depletion. Plants and animals are harmed as a result of stratospheric ozone depletion and increased solar UV radiation. Because sunlight is required for photosynthesis in terrestrial plants, they are most vulnerable to UV radiation [14, 40].

5.8 Herbicide toxicity

Herbicides are frequently used to control weeds in cultivated agricultural plants. Nonetheless, unintentional pesticide usage may cause oxidative stress in plants. Herbicides cause oxidative stress by producing too much ROS, which degrades plant photosynthetic pigments, lipids, cell membranes, and enzyme activity, affecting plant growth and production. Glyphosate, an herbicide, generated this stress in plants by limiting the shikimate pathway, resulting in excessive production of ROS and disruption of redox homeostasis. Glyphosate strongly hindered the growth of *Hordeum vulgare* L. (Barley) in response to larger accumulations of H_2O_2 and O_2, which raised LPO [41, 42].

6. ROS defense machinery

This defense mechanism is comprised of antioxidant machinery, which aids in the mitigation of the above-mentioned oxidative stress-induced harms. The antioxidant machinery is composed of two arms: enzymatic components and nonenzymatic antioxidants (**Figure 2**) [43].

6.1 Enzymatic antioxidants

Within the cellular and subcellular compartments, the antioxidant machinery is composed of several antioxidant enzymes such as superoxide dismutase (SOD), ascorbate peroxidase (APX), peroxidase (POD), catalase (CAT), dehydroascorbate reductase (DHAR), guaiacol peroxidase (GPX), glutathione reductase (GR), and monodehydroascorbate reductase (MDHAR).

6.1.1 Superoxide dismutase (SOD)

All aerobic organisms contain the metalloenzyme SOD. This is the initial line of defense against ROS-induced damage under environmental stress. SOD catalyzes $O_2^{\bullet-}$ elimination by dismutating it into H_2O_2 and O_2. This eliminates $OH^{\bullet}$ generation as a result of the Haber-Weiss reaction. The SODs can be classified into three groups based on the metal ion they bind: Mn-SOD located in mitochondria, Cu/Zn-SOD located in cytosol, (chloroplasts and peroxisomes), and Fe-SOD localized in the chloroplasts. Under abiotic stress (including salinity), the SOD levels get increased, as plants tend to defend themselves from oxidative stress [21].

$$O_2^{\bullet-} + O_2^{\bullet-} + 2H^- \rightarrow 2H_2O_2 + O_2 \quad (1)$$

6.1.2 Catalase (CAT)

It is a tetrameric heme enzyme that catalyzes the dismutation of H_2O_2 into O_2 and H_2O. It has a strong affinity for H_2O_2, but a weaker attraction for organic

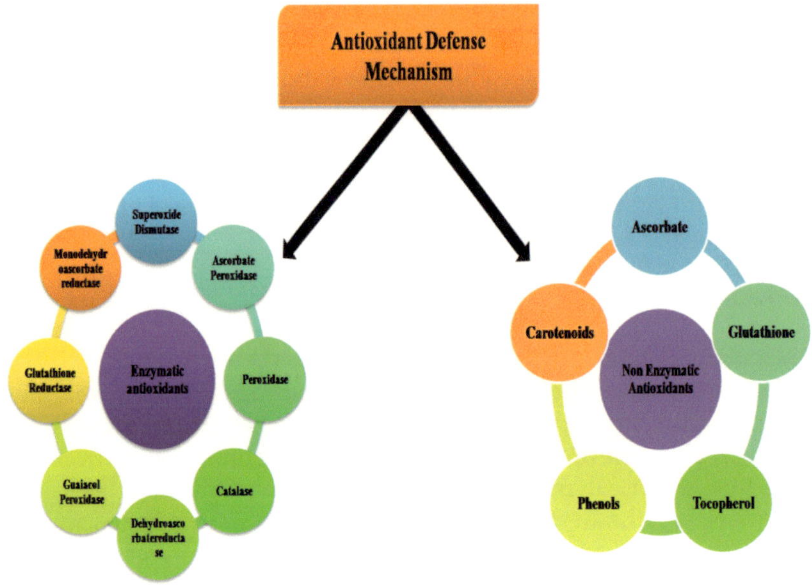

Figure 2.
Components of antioxidant defense mechanism.

peroxides. It has a very rapid turnover rate and is unique among antioxidant enzymes in that it does not require a reducing counterpart. Peroxisomes are hotspots for H_2O_2 generation due to photorespiration, purine catabolism, fatty acid oxidation, and oxidative stress. Recent studies reveal that CAT is also located in other subcellular compartments including chloroplast, mitochondria, and cytosol; however, considerable CAT activity has yet to be observed. CAT eliminates H_2O_2 in an energy-efficient manner [21, 44]:

$$2H_2O_2 \rightarrow H_2O + {}_{(1/2)}O_2 \quad (2)$$

6.1.3 Ascorbate peroxidase (APX)

The ascorbate glutathione cycle (ASC-GSH) is driven by ascorbate peroxidase (APX). While CAT primarily scavenges H_2O_2 in the peroxisomes, APX does the same in the cytosol and chloroplast. The APX uses ascorbic acid (AA) as a reducing agent to convert H_2O_2 to H_2O and DHA.

$$H_2O_2 + AA \rightarrow 2H_2O + DHA \quad (3)$$

The APX family is divided into five distinct isoforms depending on amino acid sequences and location, including chloroplastid (thylakoidal and stromal), mitochondrial, peroxisomal, and cytosolic. Because APX is more extensively distributed and has a higher affinity toward H_2O_2 than CAT, it is a more effective H_2O_2 scavenger during stress [45].

6.1.4 Monodehydroascorbate reductase (MDHAR)

It is in charge of regenerating AA from the short-lived MDHA by employing NADPH as a reducing agent and thus refilling the cellular AA pool. Because it

DOI: http://dx.doi.org/10.5772/intechopen.101045

regenerates AA, it colocalizes with APX in mitochondria and peroxisomes where APX scavenges H_2O_2 and oxidizes AA. MDHAR has several isozymes that are found in mitochondria, peroxisomes, chloroplasts, cytosol, and glyoxysomes [2, 46]:

$$\text{MDHA} + \text{NADPH} \rightarrow \text{AA} + \text{NADP}^{+} \quad (4)$$

6.1.5 *Dehydroascorbate reductase (DHAR)*

It uses reduced glutathione (GSH) as an e^- donor to convert dehydroascorbate (DHA) to AA. This makes it, in addition to MDHAR, another drug that regenerates the cellular AA pool. It is essential for regulating the AA pool size in both apoplast and symplast, hence maintaining the plant cell's redox status. DHAR can be detected in high concentrations in seeds, roots, and both green and etiolated shoots [47, 48].

$$\text{DHA} + 2\text{GSH} \rightarrow \text{AA} + \text{GSSG} \quad (5)$$

6.1.6 *Glutathione reductase (GR)*

It is a flavoprotein oxidoreductase that reduces GSSG to GSH using NADPH as a reductant. Reduced glutathione (GSH) is depleted when it is utilized to produce AA from DHA and MDHA, and it is thus transformed to its oxidized state (GSSG). To maintain a high cellular GSH/GSSG ratio, GR, a critical enzyme of the ASC-GSH cycle, catalyzes the creation of a disulfide bond in glutathione disulfide. It is mostly found in chloroplasts, with trace amounts being detected in cytoplasm and mitochondria. GSH is a low-molecular-weight molecule that acts as a reductant, preventing thiol groups from being oxidized and reacting with harmful ROS members [49, 50].

$$\text{GSSG} + \text{NADPH} \rightarrow 2\text{GSH} + \text{NADP}^{+} \quad (6)$$

6.2 Nonenzymatic components of antioxidative defense system

The primary cellular redox buffers glutathione, ascorbate, phenols, carotenoids, and tocopherol that are nonenzymatic components of the antioxidative defense mechanism. Plant growth and development are influenced by numerous antioxidants that modulate several processes ranging from cell elongation and mitosis to senescence and cellular death. They not only protect various cell components from harm, but also play a significant role in plant growth, maturation, and development by modifying cellular processes such as mitosis, senescence, cell elongation, and cell death.

6.2.1 *Ascorbate (AsA)*

AsA is the most prevalent, low-molecular-weight antioxidant, and it plays an important role in the defense against oxidative stress induced by increased ROS levels. Because of its potential to donate e^- in a variety of enzymatic and nonenzymatic processes, AsA is regarded as a potent antioxidant. AsA has been demonstrated to be involved in several physiological processes in plants. The majority of AsA, about 90%, is found in the cytoplasm, but unlike other soluble antioxidants, a significant fraction is transported to the apoplast. The first line of defense in apoplastic AsA is thought to be against potentially harmful exogenous oxidants. AsA shields important macromolecules from oxidative degradation. Under normal physiological conditions, AsA is usually found in a reduced state in chloroplasts. It protects the membrane by directly interacting with H_2O_2, $O_2^{\bullet}$, producing tocopherol from tocopheroxyl radicals, and preserving the activities of enzymes containing prosthetic transition metal ions. AsA plays a beneficial role in the elimination of H_2O_2 *via* the AsAGSH cycle [51, 52].

6.2.2 Glutathione

Glutathione tripeptide (−glutamylcysteinyl-glycine, GSH) is a critical low-molecular-weight nonprotein thiol that plays a significant role in intracellular defense action against ROS-induced oxidative damage. It has been documented in almost every cell compartment, including the chloroplasts, cytosol, vacuoles, endoplasmic reticulum, and mitochondria. GSH is produced in plant cells, chloroplasts, and cytosol, by compartment-specific isoforms of glutathione synthetase and glutamylcysteinyl synthetase. The balance of glutathione disulfide (GSSG) and GSH is essential for sustaining cellular redox state. A variety of biological processes depend on GSH's reducing abilities, including cell growth/division, sulfate transport regulation, metabolite conjugation, signal transduction, enzymatic regulation, nucleic acid and protein synthesis, xenobiotic detoxification, synthesis of phytochelatins, and stress-responsive gene expression. GSH acts as an antioxidant in several ways. It can chemically react with $OH^{\bullet}$, $O_2^{\bullet -}$, and H_2O_2 and hence work directly as a free radical scavenger. GSH can protect macromolecules (proteins, lipids, and DNA) by directly forming adducts with reactive electrophiles [53].

6.2.3 Tocopherols

Tocopherols are a class of lipophilic antioxidants that scavenge oxygen-free radicals, lipid peroxy radicals, and $^1O_2$. The relative antioxidant activities of the tocopherol isomers α-, β-, γ-, and δ- are related to the number of methyl groups and methylation pattern that are connected to the phenolic ring of the polar head structure. As a result, tocopherol has the highest antioxidant activity due to its three methyl substituents. Tocopherols are only generated by photosynthetic organisms and are found only in the green portions of plants. As precursors, the tocopherol biosynthesis pathway uses two compounds: homogentisic acid (HGA) and phytyl diphosphate (PDP). Tocopherols inhibit the chain propagation stage in lipid autoxidation, making them an efficient free radical trap. In redox interactions with $^1O_2$, tocopherol's fully substituted benzoquinone ring and fully reduced phytyl chain operate as antioxidants [54].

6.2.4 Carotenoids

Carotenoids are lipophilic antioxidants capable of detoxifying several kinds of ROS. Carotenoids can be found in both plants and microbes. During the visible range between 400 and 550 nm, carotenoids absorb light and transmit it to the chloroplast. They act as an antioxidant by scavenging $^1O_2$ to avoid oxidative damage and quenching triplet sensitizer (3Chl) and excited chlorophyll (Chl) molecules to prevent the creation of $^1O_2$ and therefore protect the photosynthetic system. Carotenoids are also the precursors of several signaling molecules, which regulate various biotic/abiotic stress responses and plant development. Carotenoids' ability to scavenge, inhibit, or reduce the formation of triplet chlorophyll may be explained by their chemical specificity. Carotenoids comprise a chain of isoprene residues with many conjugated double bonds, allowing for simple energy intake from excited molecules and heat dissipation [55].

6.2.5 Phenolic compounds

These are the group of secondary metabolites with antioxidant capabilities that include tannins, flavonoids, lignin, and hydroxycinnamate esters. They are abundant in plant tissues. Polyphenols have an aromatic ring with –OH or OCH_3 substituents that contribute to their biological activity, which includes an antioxidant activity. In addition to chelating transition metal ions, polyphenols absorb molecular species of active oxygen and inhibit lipid peroxidation by scavenging lipid alkoxyl radicals. They also change the lipid packing order and reduce membrane fluidity. These modifications may severely hamper free radical transport and limit peroxidative processes. Furthermore, it has been demonstrated that flavonoids and phenylpropanoids, in particular, are oxidized by peroxidase and act in an H_2O_2-scavenging, phenolic system [56].

6.2.6 Proline

Proline, an osmolyte, is also considered a potent antioxidant. It is frequently employed as a nonenzymatic antioxidant across various kingdoms to combat the detrimental effects of various ROS members. This is produced from glutamic acid *via* a pyrroline 5-carboxylate intermediate. This route is mediated in plants by two enzymes, pyrroline-5-carboxylate reductase and 1-pyrroline-5-carboxylate synthetase. It is an effective $OH^{\bullet}$ and $^1O_2$ scavenger and can prevent LPO damage. Proline accumulates in huge concentrations in plants during stress, either due to increased synthesis or due to decreased breakdown [57].

7. Conclusion

ROS have long been known to play an important role in controlling plant responses to both biotic and abiotic stressors. They are unavoidable harmful metabolic byproducts that serve as signaling molecules under stress situations. Although ROS appear to be damage agents in plants, their importance in boosting the stress signaling component to prevent future losses is also noteworthy. Despite the constant increase in stress-related publications, there is no novelty in the content. ROS has a dual purpose: it is an unavoidable byproduct of aerobic metabolism, on the one hand, and it serves as a marker during stressful conditions, on the other hand. They not only act as damage agents in plants but also activate stress-signaling components to prevent future harm. ROS synthesis is extensive, with production sites found both intracellularly and extracellularly. ROS causes severe damage, and its targets include all biomolecules such as DNA, proteins, and lipids compromising the cell's integrity and ultimately leading to death. This chapter explains how both components of the antioxidant machinery, antioxidant enzymes and non-antioxidant metabolites work together to mitigate the harmful effects of ROS and build a tolerance to diverse environmental stress situations. Despite substantial advances in recent years, there are still uncertainties and gaps in our knowledge of ROS production and how they influence plants, owing to their short half-life and highly reactive nature. Although the highly compartmentalized nature of antioxidants is well understood, the sensing and response mechanisms, as well as the regulation of the delicate balance between production and scavenging, require more investigation.

In the future, molecular research might lead to a better understanding of ROS metabolism. Advanced functional genomics, in conjunction with proteomics and metabolomics, will provide extensive insights into the ROS network and its associated reactions.

Acknowledgements

The authors are grateful to DST FIST and Department of Bioscience, Integral University, Lucknow, Uttar Pradesh, India.

Conflict of interest

The authors declare no conflict of interest.

Author details

Ambreen Bano[1*], Anmol Gupta[1], Smita Rai[1], Touseef Fatima[1,3], Swati Sharma[1] and Neelam Pathak[2]

1 IIRC-3, Plant-Microbe Interaction and Molecular Immunology Laboratory, Department of Biosciences, Faculty of Sciences, Integral University, Lucknow, UP, India

2 Department of Biochemistry, Dr. Rammanohar Lohia Avadh University, Ayodhya, UP, India

3 CSIR-National Botanical Research Institute, Lucknow, India

*Address all correspondence to: ambreenbano2408@gmail.com

References

[1] Foyer CH, Noctor G. Redox regulation in photosynthetic organisms: Signaling, acclimation, and practical implications. Antioxidants & Redox Signaling. 2009;**11**(4):861-905

[2] Kapoor D, Bhardwaj S, Landi M, Sharma A, Ramakrishnan M, Sharma A. The impact of drought in plant metabolism: How to exploit tolerance mechanisms to increase crop production. Applied Sciences. 2020; **10**(16):5692

[3] Suzuki N, Koussevitzky SH, Mittler RO, Miller GA. ROS and redox signalling in the response of plants to abiotic stress. Plant, Cell & Environment. 2012;**35**(2):259-270

[4] Sachdev S, Ansari SA, Ansari MI, Fujita M, Hasanuzzaman M. Abiotic stress and reactive oxygen species: Generation, signaling, and defense mechanisms. Antioxidants. 2021; **10**(2):277

[5] Hasanuzzaman M, Bhuyan MH, Zulfiqar F, Raza A, Mohsin SM, Mahmud JA, et al. Reactive oxygen species and antioxidant defense in plants under abiotic stress: Revisiting the crucial role of a universal defense regulator. Antioxidants. 2020;**9**(8):681

[6] Mehla N, Sindhi V, Josula D, Bisht P, Wani SH. An introduction to antioxidants and their roles in plant stress tolerance. In: Reactive Oxygen Species and Antioxidant Systems in Plants: Role and Regulation under Abiotic Stress. Singapore: Springer; 2017. pp. 1-23

[7] Singh A, Kumar A, Yadav S, Singh IK. Reactive oxygen species-mediated signaling during abiotic stress. Plant Gene. 2019;**18**:100173

[8] Raja V, Majeed U, Kang H, Andrabi KI, John R. Abiotic stress: Interplay between ROS, hormones and MAPKs. Environmental and Experimental Botany. 2017;**137**:142-157

[9] Kuluev B, Mikhaylova E, Berezhneva Z, Nikonorov Y, Postrigan B, Kudoyarova G, et al. Expression profiles and hormonal regulation of tobacco NtEXGT gene and its involvement in abiotic stress response. Plant Physiology and Biochemistry. 2017;**111**:203-215

[10] Halliwell B, Gutteridge JM. Free Radicals in Biology and Medicine. USA: Oxford University Press; 2015

[11] Krieger-Liszkay A, Fufezan C, Trebst A. Singlet oxygen production in photosystem II and related protection mechanism. Photosynthesis Research. 2008;**98**(1):551-564

[12] Noctor G, Veljovic-Jovanovic SO, Driscoll S, Novitskaya L, Foyer CH. Drought and oxidative load in the leaves of C3 plants: a predominant role for photorespiration? Annals of Botany. 2002;**89**(7):841-850

[13] Pint BA. Experimental observations in support of the dynamic-segregation theory to explain the reactive-element effect. Oxidation of Metals. 1996; **45**(1):1-37

[14] Miller GA, Suzuki N, Ciftci-Yilmaz SU, Mittler RO. Reactive oxygen species homeostasis and signalling during drought and salinity stresses. Plant, Cell & Environment. 2010;**33**(4): 453-467

[15] Mignolet-Spruyt L, Xu E, Idänheimo N, Hoeberichts FA, Mühlenbock P, Brosché M, et al. Spreading the news: Subcellular and organellar reactive oxygen species production and signalling. Journal of Experimental Botany. 2016;**67**(13): 3831-3844

[16] Noctor G, De Paepe R, Foyer CH. Mitochondrial redox biology and homeostasis in plants. Trends in Plant Science. 2007;**12**(3):125-134

[17] Pastore D, Trono D, Laus MN, Di Fonzo N, Flagella Z. Possible plant mitochondria involvement in cell adaptation to drought stress: A case study: Durum wheat mitochondria. Journal of Experimental Botany. 2007; **58**(2):195-210

[18] Hu X, Zhang A, Zhang J, Jiang M. Abscisic acid is a key inducer of hydrogen peroxide production in leaves of maize plants exposed to water stress. Plant and Cell Physiology. 2006;**47**(11): 1484-1495

[19] Eaton CJ, Jourdain I, Foster SJ, Hyams JS, Scott B. Functional analysis of a fungal endophyte stress-activated MAP kinase. Current Genetics. 2008; **53**(3):163-174

[20] Higuchi T. Look back over the studies of lignin biochemistry. Journal of Wood Science. 2006;**52**(1):2-8

[21] Mittler R. Oxidative stress, antioxidants and stress tolerance. Trends in Plant Science. 2002;**7**(9): 405-410

[22] Yadu B, Chandrakar V, Keshavkant S. Responses of plants to fluoride: an overview of oxidative stress and defense mechanisms. Fluoride. 2016;**49**(3):293

[23] Hu WH, Song XS, Shi K, Xia XJ, Zhou YH, Yu JQ. Changes in electron transport, superoxide dismutase and ascorbate peroxidase isoenzymes in chloroplasts and mitochondria of cucumber leaves as influenced by chilling. Photosynthetica. 2008;**46**(4): 581-588

[24] Moller IM, Kristensen BK. Protein oxidation in plant mitochondria as a stress indicator. Photochemical & Photobiological Sciences. 2004;**3**(8): 730-735

[25] Baker A, Graham IA, editors. Plant Peroxisomes: Biochemistry, Cell Biology and Biotechnological Applications. Netherlands: Springer Science & Business Media; 2013

[26] Yan J, Tsuichihara N, Etoh T, Iwai S. Reactive oxygen species and nitric oxide are involved in ABA inhibition of stomatal opening. Plant, Cell & Environment. 2007;**30**(10):1320-1325

[27] Negrao S, Schmöckel SM, Tester M. Evaluating physiological responses of plants to salinity stress. Annals of Botany. 2017;**119**(1):1-1

[28] Gupta A, Rai S, Bano A, Khanam A, Sharma S, Pathak N. Comparative Evaluation of Different Salt-Tolerant Plant Growth-Promoting Bacterial Isolates in Mitigating the Induced Adverse Effect of Salinity in Pisum sativum. 2021;**11**

[29] Chourasia KN, Lal MK, Tiwari RK, Dev D, Kardile HB, Patil VU, et al. Salinity stress in Potato: Understanding physiological, biochemical and molecular responses. Life. 2021;**11**(6):545

[30] Sperdouli I, Moustakas M. Interaction of proline, sugars, and anthocyanins during photosynthetic acclimation of *Arabidopsis thaliana* to drought stress. Journal of Plant Physiology. 2012;**169**(6):577-585

[31] Cruz de Carvalho MH. Drought stress and reactive oxygen species: Production, scavenging and signaling. Plant Signaling & Behavior. 2008;**3**(3): 156-165

[32] Bhaduri AM, Fulekar MH. Antioxidant enzyme responses of plants to heavy metal stress. Reviews in Environmental Science and Bio/ Technology. 2012;**11**(1):55-69

[33] Ashraf MA. Waterlogging stress in plants: A review. African Journal of Agricultural Research. 2012;**7**(13): 1976-1981

[34] Patterson BD, Payne LA, Chen YZ, Graham D. An inhibitor of catalase induced by cold in chilling-sensitive plants. Plant Physiology. 1984;**76**(4): 1014-1018

[35] Nahar K, Hasanuzzaman M, Alam M, Fujita M. Exogenous spermidine alleviates low temperature injury in mung bean (*Vigna radiata* L.) seedlings by modulating ascorbate-glutathione and glyoxalase pathway. International Journal of Molecular Sciences. 2015;**16**(12):30117-30132

[36] Dar MA, Reshi ZA. Induced genotoxicity and oxidative stress in plants: An overview. Induced Genotoxicity and Oxidative Stress in Plants. 2021;**1**:1-27

[37] Nahar K, Hasanuzzaman M. Germination, growth, nodulation and yield performance of three mungbean varieties under different levels of salinity stress. Green Farming. 2009; **2**(12):825-829

[38] Zhu WJ, Li PP, Xue CM, Chen M, Wang Z, Yang Q. Potato plants overexpressing SpHsfA4c exhibit enhanced tolerance to high-temperature stress Russian Journal of Plant Physiology 2021:1-0

[39] Sonkar S, Singh AK, Husen A. Functions of hydrogen sulfide in plant regulation and response to abiotic stress. In: Plant Performance Under Environmental Stress. Champions: Springer; 2021. pp. 329-355

[40] Jeena AS, Pandey D. Metal induced genotoxicity and oxidative stress in plants, assessment methods, and role of various factors in genotoxicity regulation. In: Induced Genotoxicity and Oxidative Stress in Plants. Singapore: Springer; 2021. pp. 133-149

[41] Bela K, Horváth E, Gallé Á, Szabados L, Tari I, Csiszár J. Plant glutathione peroxidases: Emerging role of the antioxidant enzymes in plant development and stress responses. Journal of Plant Physiology. 2015;**176**: 192-201

[42] Caverzan A, Piasecki C, Chavarria G, Stewart CN, Vargas L. Defenses against ROS in crops and weeds: The effects of interference and herbicides. International Journal of Molecular Sciences. 2019; **20**(5):1086

[43] Bano A, Gupta A, Rai S, Sharma S, Pathak N. Elucidation of bioactive potential of two commonly grown north Indian *Psidium guajava* viz. Lalit and Shweta against Pathogenic Foodborne and MDR Bacteria. 2021;**11**(6):14090-14102

[44] Zimmermann PE, Zentgraf UL. The correlation between oxidative stress and leaf senescence during plant development. Cellular and Molecular Biology Letters. 2005;**10**(3):515

[45] Sharma P, Dubey RS. Ascorbate peroxidase from rice seedlings: properties of enzyme isoforms, effects of stresses and protective roles of osmolytes. Plant Science. 2004;**167**(3): 541-550

[46] Stevens R, Page D, Gouble B, Garchery C, Zamir D, Causse M. Tomato fruit ascorbic acid content is linked with monodehydroascorbate reductase activity and tolerance to chilling stress. Plant, Cell & Environment. 2008;**31**(8): 1086-1096

[47] Eltayeb AE, Kawano N, Badawi GH, Kaminaka H, Sanekata T, Shibahara T, et al. Overexpression of monodehydroascorbate reductase in transgenic tobacco confers enhanced tolerance to ozone, salt and polyethylene glycol stresses. Planta. 2007;**225**(5): 1255-1264

[48] Chen Z, Gallie DR. Dehydroascorbate reductase affects leaf growth, development, and function. Plant Physiology. 2006;**142**(2):775-787

[49] Karuppanapandian T, Moon JC, Kim C, Manoharan K, Kim W. Reactive oxygen species in plants: Their generation, signal transduction, and scavenging mechanisms. Australian Journal of Crop Science. 2011;**5**(6): 709-725

[50] Gill SS, Anjum NA, Hasanuzzaman M, Gill R, Trivedi DK, Ahmad I, et al. Glutathione and glutathione reductase: A boon in disguise for plant abiotic stress defense operations. Plant Physiology and Biochemistry. 2013;**70**:204-212

[51] Smirnoff N. Ascorbic acid: Metabolism and functions of a multi-facetted molecule. Current Opinion in Plant Biology. 2000;**3**(3):229-235

[52] Barnes J, Zheng Y, Lyons T. Plant resistance to ozone: The role of ascorbate. In: Air Pollution and Plant Biotechnology. Tokyo: Springer; 2002. pp. 235-252

[53] Foyer CH, Lopez-Delgado H, Dat JF, Scott IM. Hydrogen peroxide- and glutathione-associated mechanisms of acclimatory stress tolerance and signalling. Physiologia Plantarum. 1997;**100**(2):241-254

[54] Kamal-Eldin A, Appelqvist LÅ. The chemistry and antioxidant properties of tocopherols and tocotrienols. Lipids. 1996;**31**(7):671-701

[55] Li F, Vallabhaneni R, Yu J, Rocheford T, Wurtzel ET. The maize phytoene synthase gene family: Overlapping roles for carotenogenesis in endosperm, photomorphogenesis, and thermal stress tolerance. Plant Physiology. 2008;**147**(3):1334-1346

[56] Hadi SM, Bhat SH, Azmi AS, Hanif S, Shamim U, Ullah MF. Oxidative breakage of cellular DNA by plant polyphenols: A putative mechanism for anticancer properties. Seminars in Cancer Biology. 2007;**17**(5):370-376

[57] Nathalie V, Christian H. Proline accumulation in plants: A review. Amino Acids. 2008;**35**(4):753-759

Chapter 7

Phenolic Compounds in the Plant Development and Defense: An Overview

Sambangi Pratyusha

Abstract

Phenolic compounds are produced by the plants mainly for their growth, development, and protection. These aromatic benzene ring compounds are very much essential during the plant's biotic and abiotic stress interactions. They constitute an essential part of plant's secondary metabolites and play a vital role in various physiological and mechanical activities. These diverse plant phenolic compounds act both as attractants and repellents toward various organisms in the environment. They could act as attractants toward the beneficial organisms and as toxicants against the invading pests and pathogens. These metabolite compounds often enhance during a plethora of stress conditions and act as the first line of defense to provide plant disease resistance. They are also known to influence the other plant metabolic pathways, namely phytoalexin biosynthesis and reactive oxygen species generation. These phenolic compounds participate both in the above- and below-ground plant defense systems. They are produced as root exudates and influence the soil diversity and the neighboring plants. The present review provides an overview of the roles of plant phenolic compounds in the plant kingdom as signaling compounds, pigment compounds, antimicrobials, and defense compounds.

Keywords: abiotic and biotic stress, pest management, plant defense, plant phenols

1. Introduction

Plants produce an amazing diversity of secondary metabolites, and the most important ones are the phenolic compounds. They are the most stable products in the plant kingdom. Humans have known them for centuries, and their role in plants' nutrition, fertility, growth, and protection has made them compounds of interest and to understand them completely. These anti-herbivore chemicals produced by plants are one of the most common plant allelochemicals in the ecosystem. These phenolic compounds are characterized by single or more hydroxyl groups bound to a six-carbon aromatic ring. These compounds attained the leading status due to the resistant properties bestowed by them [1]. More than 8000 phenolic structures are currently known, ranging from simple phenolic acids to highly polymerized substances as tannins [2]. They are the most abundant secondary metabolites with wide distribution in the plant kingdom.

Primarily these phenolic compounds are usually involved in the plant defense responses and apart from that, they are were also seen playing a role during crop pollination and camouflage [3, 4]. These compounds are mostly found bound to the sugars. Most of them being aromatic in nature also play an important role in plant communication. At the plant's rhizosphere, certain phenolic compounds monitor their surroundings through quorum sensing [5]. The plant growth-promoting microbes at the rhizosphere breakdown these phenolics and, in turn, enhance the soil fertility. They also aid in the chelation of the soil minerals and elements, improve the soil porosity, and in turn increase the absorption capacity of the plants [6]. During stress and pathogen invasion, these phenolic compounds are often accumulated in the plant's subepidermal tissues. The synthesis and concentration of the accumulated phenolics depend on many internal and external factors such as plant physiology, age, development stage, climate, and the type of pathogen attack [7]. The significant nature of the phenolic compounds is their dual function as both attractants and repellent compounds. Depending on the surrounding environment, the plant produces either the attractants phenolic derivatives, namely allelochemicals and chemoattractants, to attract the pollinators, symbiotic microbes [8, 9], or repellent phenolic derivatives to repel the pests and pathogens [10].

Due to present-day environmental challenges, there is a need for eco-friendly agricultural practices, and to ensure the future demand for food, exploitation of sustainable solutions is very much necessary. This brings us to the attention of plant phenolic compounds with diverse beneficial roles as plant growth promoters, crop yield enhancers, and as protectors against varied environmental stresses. Hence, the present chapter explores the diversity of plant phenolic compounds and their role in plant development and defense toward their application in various fields of agriculture.

2. Classification of phenolic compounds

According to Harborne [2], these plant phenolic compounds are mainly classified (**Table 1**) into the following groups.

Number of C atoms	Basic skeleton	Class
6	C_6	Simple phenols, benzoquinones
7	$C_6–C_1$	Phenolic acids
8	$C_6–C_2$	Acetophenone, phenylacetic acid
9	$C_6–C_3$	Hydroxycinnamic acid, polypropene, coumarin, isocoumarin
10	$C_6–C_4$	Naphthoquinone
13	$C_6–C_1–C_6$	Xanthone
14	$C_6–C_2–C_6$	Stilbene, anthrachinone
15	$C_6–C_3–C_6$	Flavonoids, isoflavonoids
18	$(C_6–C_3)_2$	Lignans, neolignans
30	$(C_6–C_3–C_6)_2$	Biflavonoids
n	$(C_6–C_3)_n$ $(C_6)_n$ $(C_6–C_3–C_6)_n$	Lignins Catecholmelanine Condensed tannins

Table 1.
Classification of phenolic compounds based on the number of carbons.

C_6-Simple phenols and benzoquinones are single benzene ringed structures with certain medicinal benefits. For example, embelin is a plant benzoquinone with antispermatogenic effect isolated from seeds of *Embelia ribes* [11].

C_6–C_1-Phenolic acids are those compounds possessing one carboxylic acid functional group. These naturally occurring phenolic acids contain two distinctive carbon frameworks, namely the hydroxycinnamic acid and hydroxybenzoic structures. The basic structure remains the same, but the number and position of hydroxyl groups differ between the two. Phenolic acids with hydroxycinnamic acid include cinnamic acid, coumaric acid, ferulic acid, sinapic acid, and caffeic acid. Structures with hydroxybenzoic acid include benzoic acid, *p*-hydroxybenzoic acid, vanillic acid, gallic acid, protocatechuic acid, syringic acid, gentisic acid, veratric acid, and salicylic acid [12]. C_6–C_2-Acetophenone is a naturally occurring phenol compound in apple, cheese, apricot, beef, and cauliflower. It is used in fragrances and chewing gum. Phenylacetic acid is an active auxin, a plant hormone found in fruits [13]. C_6–C_3-Coumarins are notably in high concentration in *Dipteryx odorata* [14] and produced by plants as a defense chemical to discourage predation. They are widely spread in the grasses and cloves. C_6–C_4-Naphthoquinones such as 2-hydroxynapthoquinone and naphthazarin show insecticidal activity against tobacco culture insects extracted from *Calceolaria andina* [15]. Derivatives of 1,4-naphthoquinone are known to possess antibacterial and antitumor properties. Naphthoquinones also exhibit larvicidal and molluscicidal activities. They are effective against *Aedes aegypti* and *Biomphalaria glabrata* [16].

C_6–C_1–C_6-Xanthones are present in Bonnetiaceae and Clusiaceae families. They are generally used as an insecticide and as ovicide for codling moth eggs [17]. C_6–C_2–C_6-Stilbenes on hydroxylation from stilbenoids acts as phytoalexins in the plant. Commonly found plant compounds with stilbene structures are trans-resveratrol, trans-piceid, pinosylvin, piceatannol, pinosylvin, trans-pterostilbene, astringin, and rhapontin [18]. Anthraquinones generally present in plants as glycosides. C_6–C_3–C_6-Flavonoids generally occur in plants as glycosylated derivatives. The basic flavonoid structure contains a flavan nucleus. Many classes of flavonoids are present such as flavones (apigenin, luteolin, chrysin), flavan-3-ols (catechin, epicatechin, epigallocatechin), flavanones (naringenin, naringin, hesperetin, hesperidin), flavonols (quercetin, kaempferol, galangin, fisetin, myricetin), flavanonol (taxifolin), isoflavones (genistein, genistin, daidzein, daidzin, ononin), and anthocyanidins (cyanidin, cyanin, peonidin, delphinidin, pelargonidin, and malvidin) [12].

$(C_6$–$C_3)_2$-Lignans are phytonutrients with antioxidant property. Examples include pinoresinol, podophyllotoxin, and steganacin. Flax and sesame seeds contain high levels of lignans. $(C_6$–C_3–$C_6)_2$-Biflavonoids formed through an oxidative coupling of two chalcone units and subsequent modification of the central C_3 units. They are characteristic of gymnosperms, the Psilotales, Selaginallales, and several flowering plants. They are not found in Pinaceae or the Gnetales. These biflavonoids act as fungitoxins and insect feeding deterrents [19]. $(C_6$–$C_3)_n$-Lignin and tannins are polymer forms of phenolic compounds. Lignin is the largest source of phenolic material in the plant cell walls. Tannins are distributed in plants as ellagitannins, condensed tannins, and gallotannins. These tannins reduce the digestibility of plants by herbivores.

The most crucial property of phenols is their antioxidant capacity. It protects the organism against reactive oxygen species (ROS). Plant polyphenols have multifunctional properties such as reducing agents, hydrogen-donating antioxidants, and singlet oxygen quenchers and flavonoids.

3. Role of phenols in plant kingdom

Phenolic compounds exist throughout the plant kingdom, and their presence varies according to the phylum. Bryophytes are the regular producers of polyphenols, including flavonoids, but it is in the vascular plants that the full range of polyphenols was found. These phenolics survived through natural selection and upgraded through ages in types and functions. The taxonomists often use them to classify and separate the plant species. Plants synthesize these phenolic compounds unlike animals because plants are stationary to escape their predators and therefore have evolved this chemical defense against predators. The primary established roles of plant phenolics are ecological, some having dual or even multiple functions (**Figure 1**). Several studies have indicated a high degree of compartmentation of phenolic compounds and of the enzymes involved in their biosynthesis that occurs through various pathways [20].

These phenolics are widely distributed in the plant. They usually accumulate in the central vacuoles of guard cells and epidermal cells and subepidermal cells of leaves and shoots. Some found covalently linked to the plant cell wall, and others occur in waxes or on the external surfaces of plant organs. According to some findings, the deposition of flavonoids in nuclei is seen in certain tree species [21]. And it leads to a flavonoid-DNA complex that provides mutual protection against oxidative damage. In

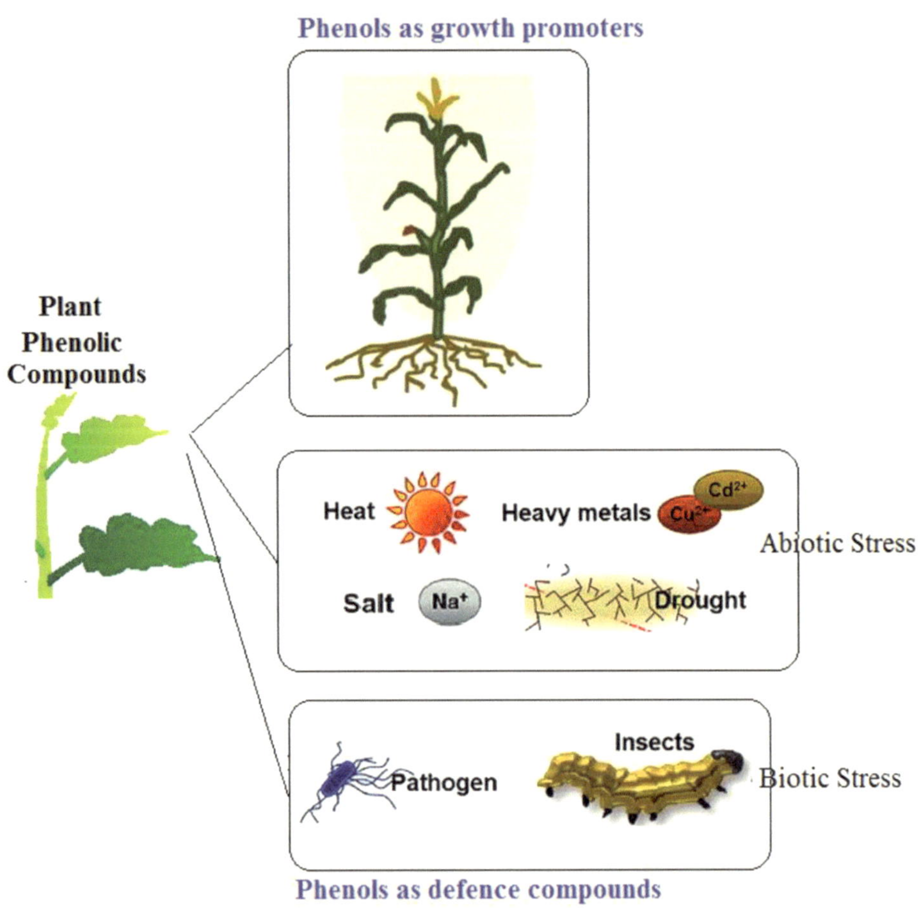

Figure 1.
Overview of functions of plant phenolic compounds.

plants, phenolics were generally produced during two scenarios such as: (1) preformed phenolics synthesized during the normal development of plant tissues and (2) induced phenolics synthesized by plants in response to physical injury, infection, and response to elicitors such as heavy metal salts, UV irradiation, temperature, etc. [22].

3.1 Role of phenols in plant growth

Majority of phenols are responsible for the growth of the plant by aiding in cell wall formation. Hydroxycinnamic acids, particularly *p*-coumaric acid and ferulic acid, were found in the insoluble or cell wall fraction as esters. These pools of wall-bound phenolic acids act as a reservoir of phenylpropanoid units for lignin biosynthesis or even that they represent the beginnings of lignification itself. These esters with a large population of bound molecules are responsible for transduction of light energy leading to changes in plant cell wall structure, water flux, turgor pressure, and growth. Auxin (indole acetic acid), a phytohormone, plays a major role in the growth regulation of the plant [23, 24].

Plants are generally rooted and unable to move from place to another and are known to move in certain ways. The circadian rhythmic leaf movement known as nyctinasty was observed in all leguminous plants. Nyctinastic leaf movement is induced by the swelling and shrinking of motor cells in the pulvini, an organ located in the joint of the leaf and believed to be controlled by Schildknecht's turgorins, which induce leaf-closing movement of the plants [25]. These turgorins are a new class of phytohormones that regulates the turgor of the plants. Some identified phenolic turgorins are: gallic acid 4-*O*-(β-D-glucopyranosyl-6′-sulfate) and gentisic acid 5-*O*-β-D-glucopyranoside that are pulvini-localized in *Mimosa pudica* L., *cis-p*-coumaric acid 4-*O*-β-D-glucopyranoside, found in *Cassia mimosoides* L., and *cis-p*-coumaroylagmatine, identified in *Albizia julibrissin* Durazz. Hence, phenols also aid in the movements of plants [26]. In rapidly germinating seeds, coumaric acid β-glucoside is more prevalent, and in non-germinating seeds, such as *Melilotus alba* found to possess a large number of free coumarins [27]. Another naturally occurring phenolic compound, which inhibits the germination of seeds, is ferulic acid. These phenolics act as germination inhibitors by inhibiting the transport of amino acids and forming of the proteins in the seeds [28].

3.2 Role of phenols in plant signaling

Allelochemicals are known to interact between two plants. Phenolic compounds influence many organic and inorganic nutrients surrounding them. They affect decomposition rate and play a role in nutrient cycle by inhibiting or stimulating spore germination. Phenols are good allelochemicals present in all parts of the plant. Leaf phenolic allelochemicals include *p*-hydroxybenzoic acid, *p*-coumaric acid, and bark, rhizosphere, root exudates including quercetin, juglone, catechin, and sorgoleone compounds [29]. Polyphenols stored in the vacuoles of plant encounter the cytoplasmatic proteins and form polyphenols-protein complex [30]. This complex aids in the senescence of plant tissues and causes the brown color of senescent leaves. Flavonoids such as eriodictyol and apigenin-7-*O*-glucoside isolated from pea root exudates found to play a role in the induction of *nod* gene expression [31]. Other natural flavanones such as naringenin, hesperitin, chalcones, and isoflavonoids such as daidzein, genistein released from legume plants inducing the *nod* gene expression. Phenols, in major flavonoids, are responsible for the pigmentation of flowers and fruits in plants and aid in the pollination and seed dispersal. For example, apigenin, luteolin, kaempferol, quercetin, and myricetin produce white, yellow, or ivory colors at their locations in plants [32].

4. Plant phenols during biotic stress

4.1 Insect-plant interactions

Among the chemical defensive strategies developed by the plant, phenolics generated due to insect herbivory play a significant role in plant resistance and controlling the herbivore damage. Many studies were performed to determine the qualitative and quantitative changes of phenolic compounds in plants and their influence on insects. Studies were conducted on rice plants infested with pests, and an elevated level of phenolic compounds was observed. The elevation was explained as a mechanism of defense that acts as a barrier to insect feeding [33, 34]. Interesting results were obtained in phenolic acid estimations by HPLC in the pest-infested plants. There was an enhancement in the level of phenolic acids in all the groundnut plants infested with the three pests compared with control plants. Also, a quantitative difference in phenolic acids was noted in the infested groundnut plants irrespective of the type of feeding damage [35]. The accumulation of the phenolic compounds by the phenylpropanoid pathway has been reported earlier [36]. Certain phenolic acids such as cinnamic acid derivatives, cinnamic acid, vanillic acid, syringic acid, and *p*-coumaric acid were found only in the pest-infested rice plants. They were totally absent in normal healthy plants. Similar results were observed with raise in the concentration of phenolic acids such as vanillic acid, syringic acid, cinnamic acid, and *p*-coumaric acid in the infested rice plants [37, 38] and cotton [39]. Kelly and Felton [40] and Rehman et al. [41] found that increased concentration in plant phenolic compounds is according to the extent of tissue damaged by feeding insects or due to pathogen infection.

Insect damage often alters plant physiology and chemistry. Larvae of the autumnal moth, *Epirrita autumnata*, on individual branches of its primary host plant, mountain birch, *Betula pubescens* did not lead to a systemic change of primary nutrients and phenol compounds. However, they affected the larval growth [42]. Changes in phloem phenols occur when pest infestation is seen on the bark of the trees. *Phytophthora ramorum*, which caused cankers on the oak trees, is analyzed for the phenolic levels against the uninfested oak tree. Ockels et al. [43], quantified nine phenolic compounds and gallic acid, tyrosol levels and showed dose-dependent inhibitory effects against *P. ramorum*, *P. cinnamomi*, *Pseudocaecilius citricola*, and *P. citrophthora* that are tested through in vitro bioassays. Seeds infested by wheat midge larvae, *Sitodiplosis mosellana*, showed induced changes in the dynamics of the phenolic acids. Analysis by HPLC of seed extracts produced by alkaline hydrolysis revealed rapid changes in *p*-coumaric and ferulic acids levels during early seed development [44]. This notified the role of phenols in seed development. Bi et al. [45] concluded that changes in the plant chemicals would induce resistance in the plant. Leaf phenolics and alkaloids variations were seen when *Coffea* spp. infested by leaf miner *Leucoptera coffeella*. The insect-plant interactions are studied by Magalhães et al. [46], to determine the pesticide activity of the plant phenols. So, sometimes the phenolic changes in specific insect-plant interaction will affect the other generalist insects of the plant.

4.2 Microbe-plant interactions

Studies indicated that microbial infection on the plant alters the plant's chemical composition. First identified phenolics providing disease resistance were seen in onion scales infected by *Colletotrichum circinans*. In order, to prevent this onion smudge disease, plant accumulated sufficient amounts of catechol and protocatechuic acid [47]. A decrease in the phenolic content of the plant was observed in brown spot infection of rice due to infection of *Helminthosporium oryzae* [48, 49]. Infection suppressed the phenol metabolism in the rice plant due to the *Helminthosporium oryzae* toxin and aided in pathogen colonization. Phenolic compounds are also involved in defense response of plants by reducing the incidence of Fusarium wilt of tomatoes caused by the fungus *Fusarium oxysporum* [50]. Alteration in the ferulic, caffeic, and vanillic acid contents and concentrations are identified from recovered leaves and roots. Elicitors from *Fusarium oxysporum* f.sp. cubense accumulated soluble and wall-bound phenolics and phenolic polymers in the roots of *Musa acuminate*. White mold fungus, *Sclerotium rolfsii* Saccodes infection to *Arachis hypogea* reduced the total soluble phenolic content [51]. However, generally phenolic compounds induce in the infected fungal plants to confer resistance to specific fungal pathogens.

According to Beckman, phenolic compounds play an important role in reducing wilt diseases and aid in signaling for the host defense responses. 4-hydroxycinnamic acid CoA ligase enzyme is vital in the diversion of phenylpropanoids, was altered by the fungi, responsible for the changes in the phenols of the infected plant. In response to Rhizobium and vesicular arbuscular mycorrhizal (VAM) inoculation, enhancement in the phenolic compounds is seen in the *Arachis hypogaea* roots [52]. These phenolic compounds released at the roots help in maintaining the Rhizobium community at the rhizosphere. Furthermore, these Rhizobia species can use the phenols as a carbon source. Bacterial pathogens also affect phenol accumulation and production. *Pseudomonas syringae* enhanced the extracellular phenolic accumulation and changed the composition of phenolic acids in the *Nicotiana tabacum* [53].

Antimicrobial activity of phenolic compounds was observed in Finnish berries against probiotic bacteria and other intestinal bacteria. Myricetin inhibited the growth of lactic acid bacteria and mostly Gram-negative bacteria [54]. Phenolic acids from *Olea europaea* leaves, namely caffeic acid, verbascoside, oleuropein, luteolin 7-*O*-glucoside, rutin, apigenin 7-*O*-glucoside, and luteolin 4′-*O*-glucoside showed antibacterial and antifungal action. Many herbs and spice plant extracts contain phenols with antibacterial activity against food-borne pathogens [55]. Flavones and flavanones of fruits and vegetables are known to be active against *Aspergillus* sp., *B. cinerea*, and *F. oxysporum* [56]. Resveratrol from grapes is also known to possess antibacterial activity [57]. In these host-microbe interactions, the phenolic metabolites play a key role in providing signals for the interactions [58]. For example, acetosyringone, a phenolic compound produced at the wound site of plants, triggers vir genes in the pathogen. In legume-rhizobial interactions, flavonoids activate the nod genes in Rhizobium responsible for symbiotic relation [59]. The roles played by these phenolic compounds generally include phytoalexins for disease defense and lignin production for structural strength, along with antioxidant nature to combat the pro-oxidants produced during microbial stress.

4.3 Effect of plant phenols on pests

Some plants respond to herbivore damage by increasing chemical, physical, or biotic defenses and responses that can help protect the remaining tissue against further damage [60–62]. Plant phenolics are believed to play an important role in chemical defense against herbivores through specific physiological effects on insects. These phenols are often described as antifeedant, digestibility reducers, and as toxins. These phenols would promote ROS in the insect digestive tract, particularly in mid-gut, where pH is alkaline. These ROS would result in direct oxidative damage to mid-gut lipids and proteins. Elevated levels of lipid per-oxidation products, oxidized protein, and free ions due to oxidative stress are generated in mid-guts of insects, leading to death [2]. Phytoalexins will disrupt the pathogen metabolism and cellular structure. Some experimental evidence includes the medicarpin by *Medicago sativa*, rishitin by Solanaceae, and camalexin by *Arabidopsis thaliana* [63]. Tannins lead to protein inactivation in insects, by binding to salivary proteins and digestive enzymes, including trypsin and chymotrypsin. Insect herbivores that ingest high amounts of tannins fail to gain weight and eventually die [64]. Lignins, which are polymer in nature, provide strong physical barrier in the form of cell walls toward herbivores. On the other hand, furanocoumarins produced in response to the herbivore attack gets activated by ultraviolet light (UV) and integrates with the DNA of insects, leading to death [65].

5. Plant phenols during abiotic stress

Plants encounter many challenges of both biotic and abiotic stress factors in nature. Abiotic stress factors include, namely drought, salinity, heat, ultraviolet, and pesticides. Nowadays, these abiotic stresses toward plants have drastically increased due to the environment's uneven climatic conditions and man-made pollution. The increase of these abiotic stresses will radically impact the growth and development of the plants and could reduce the overall crop yield [66]. Plants to combat these abiotic stress conditions will produce a plethora of defense metabolites [67, 68]. Among them, the plant phenolic compounds are playing a vital role in coping with the abiotic stresses. Under stressful conditions, these phenolics are drastically accumulated in the plant for survival [20, 69]. Phenolic compounds, namely esters, flavonoids, lignin, and tannins, act as antioxidants and fight against these abiotic stress conditions in the plant cells [70].

In certain leafy vegetables, the salinity conditions caused an increase in the phenolic compounds to counterattack the high salt levels in the soil [71]. This increase of phenols assists in the balancing of the mineral composition in the leaves. Heavy metals in the soil also dramatically impact the physiology and metabolic activities of the plant. During such stress conditions, it has been observed that the plant flavonoids are playing a vital defense role by chelating the heavy metals [72, 73]. Climate change is one of the significant factors affecting plant growth and development. Due to adverse climatic conditions, water stress in plants has become a serious concern. Due to the lack of rainfall, drought stress is common environmental stress in many cultivated areas, and crop yield is majorly dependent on it. Under drought stress, it has been observed that plants are producing polyphenols to cope with these stress conditions by controlling the water ions flux [74, 75]. Phenolic compounds also respond to nutrition stress, cold stress, and radiation, thus providing resistance to the plant [76–78]. Salinity is a significant stress factor that limits crop yield in many areas. Under these extreme conditions, plants adapt to stress through altered metabolic pathways (**Figure 2**). For example, in a medicinal plant, *Salvia mirzayanii*, total phenolic content was increased with higher salt levels [79]. Chlorogenic acid, caffeic acid, ellagic acid, ferulic acid, gallic acid, syringic acid, vanillic acid, and *p*-coumaric acid were enhanced in *Aegilops* spp. due to salinity conditions [80]. This trend in increase of phenolic contents was reported in many plants under salt stress [81, 82]. This increase of phenols is a tolerance mechanism to maintain redox homeostasis and improve plant health. Similarly, during drought stress, to avoid oxidative damage, plants produce various phenolic compounds. These flavonoids or phenols inhibit further loss of water in the plants through the closure of stomata [83]. Many reports support the production mechanism of phenols during drought stress in plants [84, 85]. The plant phenols are the main accumulator compounds during heavy metal stress. This tendency is for the chelation of toxic metals by phenolic compounds through their carboxyl and hydroxyl groups, which participate in the chelation of the metals [78]. Plant

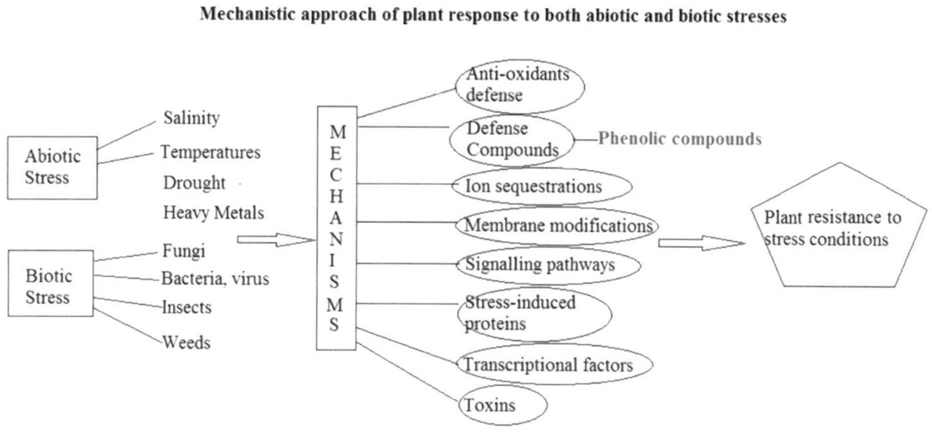

Figure 2.
Mechanistic approach of plant responses to stress conditions.

hormones play a vital role during the stress conditions. A wide cross talk of hormones, namely salicylic acid and jasmonic acid, takes place as defense response to the stresses [86]. Proteins associated with these hormones upregulate during defense and provide immunity to the plants through expression of pathogenesis-related genes [87]. Abiotic stresses significantly alter the crop quality and productivity, and to combat it, expression of resistance genes takes place and elevates the levels of defense compounds such as phenolic compounds [88]. These enhanced phenols ensure their endurance and survival of plants during these abiotic stress conditions.

6. Conclusion

Phenolic substances are the most resistant metabolites produced by plants. Better understanding of plant phenolics is essential, due to its wide array of functions in the plant development, and its practical applications in many streams such as agriculture, medicine, nutrition, pesticide management, and industry. These phenolic changes in the plant with respect to the herbivory correlated negatively with the larval growth, development, and survival of progeny. The enhanced phenols in plants behave as toxins toward insect feeding, microbial growth, and a mode of induced defense generated by the plant to defend against natural pests. We can summarize that the plant-insect interactions altered the phenolic levels to the pest attacking. So, a change in the phenol level is a defense strategy developed to combat the pest. This change of phytochemical composition from nature will oppose the pest from invading the plant. During abiotic stress also, the plants can produce phenols as tolerance mechanism to cope with the unfavorable conditions. Increased biosynthesis of plant phenols was observed in the plant during abiotic stress factors such as drought, heavy metal stress, salinity, and radiation. In conclusion, this review on the plant phenolic compounds and their role in the plant's growth, development, and defense will provide the information to understand the plant mechanisms and aid us in further effective application of them in agricultural pest management strategies.

Acknowledgements

The author thanks Department of Science & Technology (DST), India, for their support.

Conflict of interest

There are no conflicts of interest.

Author details

Sambangi Pratyusha[1,2]

1 CSIR-Indian Institute of Chemical Technology, Hyderabad, India

2 International Crops Research for Semi-Arid Tropics (ICRISAT), Hyderabad, India

*Address all correspondence to: s.pratyusha@cgiar.org

References

[1] Beninger CW, Abou-Zaid M, Kistner AL, Hallett RH, Iqbal MJ, Grodzinski B, et al. A flavanone and two phenolic acids from *Chrysanthemum morifolium* with phytotoxic and insect growth regulating activity. Journal of Chemical Ecology. 2004;**30**:589-606. DOI: 10.1023/b:joec.0000018631.67394.e5

[2] Harborne JB. General procedures and measurement of total phenolics. In: Dey PM, Harborne JB, editors. Plant Phenolics. London, UK: Academic Press; 1989. pp. 1-28. DOI: 10.1002/pca.2800020110

[3] Acamovic T, Brooker JD. Biochemistry of plant secondary metabolites and their effects in animals. The Proceedings of the Nutrition Society. 2005;**64**:403-412. DOI: 10.1079/PNS2005449

[4] Alasalvar C, Grigor JM, Zhang DL, Quantick PC, Shahidi F. Comparison of volatiles, phenolics, sugars, antioxidant vitamins, and sensory quality of different colored carrot varieties. Journal of Agricultural and Food Chemistry. 2001;**49**:1410-1416. DOI: 10.1021/jf000595h

[5] Bjarnsholt T, Givskov M. Quorum-sensing blockade as a strategy for enhancing host defenses against bacterial pathogens. Philosophical Transactions of the Royal Society B. 2007;**362**:1213-1222. DOI: 10.1098/rstb.2007.2046

[6] Bais HP, Weir TL, Perry LG, Gilroy S, Jorge M. The role of root exudates in rhizosphere interactions with plants and other organisms. Annual Review of Plant Biology. 2006;**57**:233-266. DOI: 10.1146/annurev.arplant.57.032905.105159

[7] Ozyigit II, Kahraman MV, Ercan O. Relation between explant age, total phenols and regeneration response of tissue cultured cotton (*Gossypium hirsutum* L). African Journal of Biotechnology. 2007;**6**:003-008

[8] Ndakidemi PA, Dakora FD. Review: Legume seed flavonoids and nitrogenous metabolites as signals and protectants in early seedling development. Functional Plant Biology. 2003;**30**:729-745. DOI: 10.1071/FP03042

[9] Vit P, Soler C, Tomás-Barberán FA. Profiles of phenolic compounds of *Apis mellifera* and *Melipona* spp. honeys from Venezuela. Zeitschrift für Lebensmittel-Untersuchung und -Forschung. 1997;**91**:571-577

[10] Lattanzio V, Lattanzio VMT, Cardinali A. Role of phenolics in the resistance mechanisms of plants against fungal pathogens and insects. In: Imperato F, editor. Phytochemistry Advances in Research. India: Research Signpost; 2006. pp. 23-67

[11] Lal B, Mishra N. Importance of *Embelia ribes*: An update. International Journal of Pharmaceutical Sciences and Research. 2013;**4**:3823-3838. DOI: 10.13040/IJPSR.0975-8232

[12] Murkovic M. Phenolic compounds. In: Encyclopedia of Food Sciences and Nutrition. USA: Academic Press; 2003. pp. 4507-4514

[13] Wightman F, Douglas L. Identification of phenylacetic acid as a natural auxin in the shoots of higher plants. Physiologia Plantarum. 2006;**55**:17-24. DOI: 10.1111/j.1399-3054.1982.tb00278.x

[14] Sarker SD, Nahar L. Progress in the chemistry of naturally occurring coumarins. In: Kinghorn AD, Falk H, Gibbons S, Kobayashi J, editors. Progress in the Chemistry of Organic

Natural Products. Vol. 106. Cham: Springer International Publishing; 2017. pp. 241-304

[15] Khambay B, Jewess P. The potential of natural naphthoquinones as the basis for a new class of pest control agents—An overview of research at IACR-Rothamsted. Crop Protection. 2000;**19**:597-601. DOI: 10.1016/S0261-2194(00)00078-8

[16] Ribeiro KA, de Carvalho CM, Molina MT, Lima EP, López-Montero E, Reys JR, et al. Activities of naphthoquinones against *Aedes aegypti* (Linnaeus, 1762) (Diptera: Culicidae), vector of dengue and *Biomphalaria glabrata* (Say, 1818), intermediate host of *Schistosoma mansoni*. Acta Tropica. 2009;**111**:44-50. DOI: 10.1016/j.actatropica.2009.02.008

[17] Monajjemi M, Azizi V, Amini S, Mollaamin F. Nanotheoretical studies on evaluation of anti-cancer potential on mangosteen plant. African Journal of Agricultural Research. 2011;**6**:4661-4670. DOI: 10.5897/AJAR11.942

[18] Chong J, Poutaraud A, Hugueney P. Metabolism and roles of stilbenes in plants. Plant Science. 2009;**177**:143-155. DOI: 10.1016/j.plantsci.2009.05.012

[19] Iwashina T. Flavonoid function and activity to other plants and organisms. Biological Sciences in Space. 2003;**17**:24-44. DOI: 10.2187/bss.17.24

[20] Lattanzio V. Phenolic compounds: Introduction. In: Ramawat K, Mérillon JM, editors. Natural Products. Berlin, Heidelberg: Springer; 2013. pp. 1543-1580. DOI: 10.1007/978-3-642-22144-6_57

[21] Ferreyra FML, Rius SP, Casati P. Flavonoids: Biosynthesis, biological functions, and biotechnological applications. Frontiers in Plant Science. 2012;**3**:222. DOI: 10.3389/fpls.2012.00222

[22] Xiao L, Carrillo J, Siemann E, Ding J. Herbivore-specific induction of indirect and direct defensive responses in leaves and roots. AoB Plants. 2019;**11**:plz003. DOI: 10.1093/aobpla/plz003

[23] Sun R, Sun XF, Wang SQ, Zhu W, Wang XY. Ester and ether linkages between hydroxycinnamic acids and lignins from wheat, rice, rye, and barley straws, maize stems, and fast-growing poplar wood. Industrial Crops and Products. 2002;**15**:179-188. DOI: 10.1016/S0926-6690(01)00112-1

[24] Antonova GF, Varaksina TN, Zheleznichenko TV, Stasova VV. Changes in phenolic acids during maturation and lignification of Scots pine xylem. Russian Journal of Developmental Biology. 2012;**43**:250-260. DOI: 10.1134/S1062360412040029

[25] Schildknecht H. Turgorins—New chemical messengers for plant behavior. Endeavour. 1984;**8**:113-117. DOI: 10.1016/0160-9327(84)90003-6

[26] Ueda M, Sawai Y, Shibazaki Y, Tashiro C, Ohnuki T, Amamura S. Leaf-opening substance in the nyctinastic plant, *Albizzia julibrissin* Durazz. Bioscience, Biotechnology, and Biochemistry. 1998:2133-2137. DOI: 10.1271/bbb.62.2133

[27] Sumere CFV, Wolf G, Teuchy H, Kint J. A new thin-layer method for phenolic substances and coumarins. Journal of Chromatography. 1965;**20**:48-60. DOI: 10.1016/s0021-9673(01)97365-0

[28] Mojzer BE, Hrncic KM, Škerget M, Knez Ž, Bren U. Polyphenols: Extraction methods, antioxidative action, bioavailability and anticarcinogenic effects. Molecules. 2016;**21**:901. DOI: 10.3390/molecules21070901

[29] Li ZH, Wang Q, Ruan X, Pan CD, Jiang DA. Phenolics and plant allelopathy. Molecules. 2010;**15**:8933-8952. DOI: 10.3390/molecules15128933

[30] da Silva FL, Escribano-Bailón MT, Alonso JJP, Rivas-Gonzalo JC, Santos-Buelga C. Anthocyanin pigments in strawberry. LWT- Food Science and Technology. 2007;**40**:374-382. DOI:10.1016/j.lwt.2005.09.018

[31] Begum AA, Leibovitch S, Migner P, Zhang F. Specific flavonoids induced nod gene expression and pre-ctivated nod genes of *Rhizobium leguminosarum* increased pea (*Pisum sativum* L.) and lentil (*Lens culinaris* L.) odulation in controlled growth chamber environments. Journal of Experimental Botany. 2001;**52**:1537-1543. DOI: 10.1093/jexbot/52.360.1537

[32] Brouillard R, Chassaing S, Isorez G, Kueny-Stotz M, Figueiredo P. The visible flavonoids or anthocyanins: From research to applications. Recent Advances in Polyphenol Research. 2010. DOI: 10.1002/9781444323375.ch1

[33] Usha Rani P, Jyothsna Y. Biochemical and enzymatic changes in rice plants as a mechanism of defense. Acta Physiologiae Plantarum. 2010;**32**:695-701. DOI: 10.1007/s11738-009-0449-2

[34] Pratyusha S, Usha Rani P. Physiological effects of resveratrol and coumaric acid on two major groundnut pests and their egg parasitoid behaviour. Archives of Insect Biochemistry and Physiology. 2016;**91**:230-245. DOI: 10.1002/arch.21320

[35] Pratyusha S, Usha Rani P. Induction of phenolic acids and metals in *Arachis hypogea* L. plants due to feeding of three lepidopteran pests. Arthropod-Plant Interactions. 2013;**7**:517-525. DOI: 10.1007/s11829-013-9263-2

[36] Dixon RA, Palva NL. Stress-induced phenyl-propanoid metabolism. Plant Cell. 1995;**7**:1085-1097. DOI: 10.1105/tpc.7.7.1085

[37] Hildebrand DF, Rodribuez JG, Brown GC, Luu KT, Volden CS. Peroxidative responses of leaves in two soybean genotypes injured by two spotted spider mites (Acari: Tetranychidae). Journal of Economic Entomology. 1986;**79**:1459-1465. DOI: 10.1093/jee/79.6.1459

[38] Usha Rani P, Jyothsna Y. Physiological changes in groundnut plants induced by pathogenic infection of *Cercosporidium personatum* Deighton. Allelopathy Journal. 2009;**23**:369-378

[39] Usha Rani P, Pratyusha S. Defensive role of *Gossypium hirsutum* L. anti-oxidative enzymes and phenolic acids in response to *Spodoptera litura* F. feeding. Journal of Asia-Pacific Entomology. 2013;**16**:131-136. DOI: 10.1016/j.aspen.2013.01.001

[40] Kelly J, Felton G. Plant phenolics as dietary antioxidants for herbivorous insects: A test with genetically modified tobacco. Journal of Chemical Ecology. 2002;**27**:2579-2597. DOI: 10.1023/a:1013691802028

[41] Rehman F, Khan FA, Badruddin SMA. Role of phenolics in plant defense against insect herbivory. In: Khemani LD, Srivastava MM, y Srivastava S, editors. Chemistry of Phytopotentials: Health, Energy and Environmental Perspectives. Heidelberg: Springer; 2012

[42] Lempa K, Agrawal AA, Salminen JP, Turunen T, Ossipov V, Ossipova S, et al. Rapid herbivore-induced changes in mountain birch phenolics and nutritive compounds and their effects on performance of the major defoliator, *Epirrita autumnata*. Journal of Chemical Ecology. 2004;**30**:303-321. DOI: 10.1023/B:JOEC.0000017979.94420.78

[43] Ockels F, Eyles A, Mcpherson B, Wood D, Bonello P. Phenolic chemistry of coast live oak response to *Phytophthora ramorum* infection. Journal of Chemical Ecology. 2007;**33**:1721-1732. DOI: 10.1007/s10886-007-9332-z

[44] Ding H, Lamb RJ, Ames N. Inducible production of phenolic acids in wheat and antibiotic resistance to *Sitodiplosis mosellana*. Journal of Chemical Ecology. 2000;**132**:591-605. DOI: 10.1023/A:1005412309735

[45] Bi JL, Murphy JB, Felton GW. Antinutritive and oxidative components as mechanisms of induced resistance in cotton to *Helicoverpa zea*. Journal of Chemical Ecology. 1997;**23**:97-117. DOI: 10.1023/B:JOEC.0000006348.62578.fd

[46] Magalhães P, Vieira J, Gonçalves L, Pacheco J, Guido L, Barros A. Isolation of phenolic compounds from hop extracts using polyvinylpolypyrrolidone: Characterization by high-performance liquid chromatography-diode array detection-electrospray tandem mass spectrometry. Journal of Chromatography. A. 2010;**1217**:3258-3268. DOI: 10.1016/j.chroma.2009.10.068

[47] Walker J, Stahmann M. Chemical nature of disease resistance in plants. Annual Review of Plant Physiology. 2003;**6**:351-366. DOI: 10.1146/annurev.pp.06.060155.002031

[48] Dallagnol LJ, Rodrigues FA, DaMatta FM, Mielli MVB, Pereira SC. Deficiency in silicon uptake effects cytological, physiological, and biochemical events in the rice-*Bipolaris oryzae* interaction. Phytopathology. 2011;**101**:92-104. DOI: 10.1094/PHYTO-04-10-0105

[49] Pandey S. Physiological and biochemical changes associated with the development of brown spot diseases in rice leaves. International Journal of Agricultural Science and Technology. 2018;**5**:69-78

[50] Orr R, Nelson PN. Impacts of soil abiotic attributes on Fusarium wilt, focusing on bananas. Applied Soil Ecology. 2008;**132**:20-33. DOI: 10.1016/j.apsoil.2018.06.019

[51] Cardoza YJ, Lait CG, Schmelz EA, Huang J, Tumlinson JH. Fungus-induced biochemical changes in peanut plants and their effect on development of beet armyworm, *Spodoptera exigua* Hhbner (Lepidoptera: Noctuidae), Larvae. Environmental Entomology. 2003;**32**:220-228

[52] Devi MC, Reddy MN. Phenolic acid metabolism of groundnut (*Arachis hypogaea* L.) plants inoculated with VAM fungus and Rhizobium. Plant Growth Regulation. 2002;**37**:151-156. DOI: 10.1023/A:1020569525965

[53] Baker C, Whitaker B, Roberts D, Mock N, Rice C, Deahl K, et al. Differential induction of redox sensitive extracellular phenolics during plant–bacterial interactions. Physiological and Molecular Plant Pathology. 2005;**66**:90-98. DOI:10.1016/j.pmpp.2005.05.002

[54] Pimiä PR, Nohynek L, Meier C, Kähkönen M, Heinonen M, Hopia A, et al. Antimicrobial properties of phenolic compounds from berries. Journal of Applied Microbiology. 2001;**90**:494-507. DOI: 10.1046/j.1365-2672.2001.01271.x

[55] Pereira AP, Ferreira ICFR, Marcelino F, Valentão P, Andrade PB, Seabra R, et al. Phenolic compounds and antimicrobial activity of olive (*Olea europaea* L. Cv. *Cobrançosa*) leaves. Molecules. 2007;**12**:1153-1162. DOI: 10.3390/12051153

[56] Bartmanska A, Watecka-Zacharska E, Tronina T, Poplonski J, Sordon S, Brzezowska E, et al. Antimicrobial properties of spent hops extracts, flavonoids isolated therefrom, and their derivatives. Molecules. 2018;**23**:2059. DOI: 10.3390/molecules23082059

[57] Ma DSL, Tan LT, Chan KG, Yap WH, Pusparajah P, Chuah LH, et al. Resvertrol—Potential antibacterial agent against foodborne pathogens.

Frontiers in Pharmacology. 2018;**9**:102. DOI: 10.3389/fphar.2018.00102

[58] Mhlongo MI, Piater LA, Madala NE, Labuschagne N, Dubery IA. The chemistry of plant–microbe interactions in the rhizosphere and the potential for metabolomics to reveal signaling related to defense priming and induced systemic resistance. Frontiers in Plant Science. 2018;**9**:112. DOI: 10.3389/fpls.2018.00112

[59] Mithöfer A. Suppression of plant defence in rhizobia-legume symbiosis. Trends in Plant Science. 2002;**7**:440-444. DOI: 10.1016/s1360-1385(02)02336-1

[60] Karban R, Baldwin IT. Induced Responses to Herbivory. Chicago, IL, USA: University of Chicago Press; 1997

[61] Moreira X, Galmán A, Francisco M, Castagneyrol B, Abdala-Roberts L. Host plant frequency and secondary metabolites are concurrently associated with insect herbivory in a dominant riparian tree. Biology Letters. 2018;**14**:20180281. DOI: 0.1098/rsbl.2018.0281

[62] Kovalikova Z, Kubes J, Skalicky M, et al. Changes in content of polyphenols and ascorbic acid in leaves of white cabbage after pest infestation. Molecules. 2019;**24**:2622. DOI: 10.3390/molecules24142622

[63] Jeandet P, Clément C, Courot E, Cordelier S. Modulation of phytoalexin biosynthesis in engineered plants for disease resistance. International Journal of Molecular Sciences. 2013;**14**:14136-14170. DOI: 10.3390/ijms140714136

[64] Prinz JF, Lucas PW. Saliva tannin interactions. Journal of Oral Rehabilitation. 2000;**27**:991-994. DOI: 10.1111/j.1365-2842.2000.00578.x

[65] War AR, Paulraj MG, Ahmad T, Buhroo AA, Hussain B, Ignacimuthu S, et al. Mechanisms of plant defense against insect herbivores. Plant Signaling & Behavior. 2012;**7**:1306-1320. DOI: 10.4161/psb.21663

[66] Mantri N, Patade V, Penna S, Ford R, Pang E. Abiotic stress responses in plants: Present and future. In: Abiotic Stress Responses in Plants. Vol. 2012. New York, NY, USA: Springer; 2012. pp. 1-19

[67] Parvaiz A, Satyawati S. Salt stress and phyto-biochemical responses of plants—A review. Plant, Soil and Environment. 2008;**54**:89-99. DOI: 10.17221/2774-PSE

[68] Ramakrishna A, Ravishankar GA. Influence of abiotic stress signals on secondary metabolites in plants. Plant Signaling & Behavior. 2011;**6**:1720-1731. DOI: 10.4161/psb.6.11.17613

[69] Sharma A, Shahzad B, Rehman A, Bhardwaj R, Landi M, Zheng B. Response of phenylpropanoid pathway and the role of polyphenols in plants under abiotic stress. Molecules. 2019;**24**:2452. DOI: 10.3390/molecules24132452

[70] Selmar D. Potential of salt and drought stress to increase pharmaceutical significant secondary compounds in plants. Landbauforschung Volkenrode. 2008;**58**:139-144

[71] Neocleous D, Koukounaras A, Siomos AS, Vasilakakis M. Assessing the salinity effects on mineral composition and nutritional quality of green and red "baby" lettuce. Journal of Food Quality. 2014;**37**:1-8. DOI: 10.1111/jfq.12066

[72] Williams R, Spencer JP, Rice-Evans C. Flavonoids: Antioxidants or signaling molecules? Free Radical Biology & Medicine. 2004;**36**:838-849. DOI: 10.1016/j.freeradbiomed.2004.01.001

[73] Mira L, Fernandez MT, Santos M, Rocha R, Florêncio MH, Jennings KR. Interactions of flavonoids with iron and copper ions: A mechanism for their

antioxidant activity. Free Radical Research. 2002;**36**:1199-1208. DOI: 10.1080/1071576021000016463

[74] González-Chavira MM, Herrera-Hernández MG, Guzmán-Maldonado H. Pons-Hernández JL. Controlled water deficit as abiotic stress factor for enhancing the phytochemical content and adding-value of crops. Scientia Horticulturae. 2018;**234**:354-360. DOI: 10.1016/j.scienta.2018.02.049

[75] Mirás-Avalos JM, Intrigliolo DS. Grape composition under abiotic constrains: Water stress and salinity. Frontiers in Plant Science. 2017;**8**:851. DOI: 10.3389/fpls.2017.00851

[76] Hattenschwiler S, Vitousek PM. The role of polyphenols in terrestrial ecosystem nutrient cycling. Trends in Ecology & Evolution. 2000;**15**:238-243. DOI: 10.1016/s0169-5347(00)01861-9

[77] Akula R, Ravishankar GA. Influence of abiotic stress signals on secondary metabolites in plants. Plant Signaling & Behavior. 2011;**6**:1720-1731. DOI: 10.4161/psb.6.11.17613

[78] Winkel-Shirley B. Biosynthesis of flavonoids and effects of stress. Current Opinion in Plant Biology. 2002;**5**:218-223. DOI: 10.1016/s1369-5266(02)00256-x

[79] Valifard M, Mohsenzadeh S, Kholdebarin B, Rowshan V. Effects of salt stress on volatile compounds, total phenolic content and antioxidant activities of *Salvia mirzayanii*. South African Journal of Botany. 2014;**93**:92-97. DOI: 10.1016/j.sajb.2014.04.002

[80] Razieh K, Ahmad A, Maibody MSAM. Polyphenols, flavonoids, and antioxidant activity involved in salt tolerance in wheat, *Aegilops cylindrica* and their amphidiploids. Frontiers in Plant Science. 2021;**12**:646221. DOI: 10.3389/fpls.2021.646221

[81] Muthukumarasamy M, Gupta SD, Pannerselvam R. Enhancement of peroxidase, polyphenol oxidase and superoxide dismutase activities by tridimefon in NaCl stressed *Raphanus sativus* L. Biologia Plantarum. 2000;**43**:317-320. DOI: 10.1023/A:1002741302485

[82] Navarro JM, Flores P, Garrido C, Martinez V. Changes in the contents of antioxidant compounds in pepper fruits at ripening stages, as affected by salinity. Food Chemistry. 2006;**96**:66-73. DOI: 10.1016/j.foodchem.2005.01.057

[83] Yadav B, Jogawat A, Rahman MS, Narayan OP. Secondary metabolites in the drought stress tolerance of crop plants: A review. Gene Reports. 2021;**23**:101040. DOI: 10.1016/j.genrep.2021.101040

[84] Gao S, Wang Y, Yu S, Huang Y, Liua H, Chena W, et al. Effects of drought stress on growth, physiology and secondary metabolites of Two Adonis species in Northeast China. Scientia Horticulturae. 2020;**259**:108795. DOI: 10.1016/j.scienta.2019.108795

[85] Król A, Amarowicz R, Weidner S. Changes in the composition of phenolic compounds and antioxidant properties of grapevine roots and leaves (*Vitis vinifera* L.) under continuous of long-term drought stress. Acta Physiologiae Plantarum. 2014;**36**:1491-1499. DOI: 10.1007/s11738-014-1526-8

[86] Jones JDG, Dangl JL. The plant immune system. Nature. 2006;**444**:323-329. DOI: 10.1038/nature05286

[87] Zipfel C. Plant pattern-recognition receptors. Trends in Immunology. 2014;**35**:345-351. DOI: 10.1016/j.it.2014.05.004

[88] Kumar M et al. Phenolics as plant protective companion against abiotic stress. In: Lone R, Shuab R, Kamili A, editors. Plant Phenolics in Sustainable Agriculture. Singapore: Springer; 2020:277-308. DOI: 10.1007/978-981-15-4890-1_12

Chapter 8

Glucosinolates and Its Role in Mitigating Abiotic and Biotic Stress in *Brassicaceae*

Parul Chowdhury

Abstract

Abiotic stresses such as increase in daily mean temperature, changed patterns of precipitation, increase in episodes of drought and floods in future are faced by the plants and pose threats to crop production and food security. Induction of secondary metabolites by several abiotic stress conditions can be helpful in the crop protection against biotic stress and can be a major link between biotic and biotic stress. Plants also face threats by injury caused by herbivores and insects that chew on plants. Plant develops, coordinates and combines defence mechanism to cope with the challenges caused by the injuries. The plant family *Brassicaceae* (or *Cruciferae*) includes some of the world's most economically important crops; especially members of the genera *Brassica L. Brassicaceae* vegetables are a good source of secondary metabolite that is Glucosinolates. Which are responsible for characteristic flavour and odour, when degraded. Glucosinolates and their degradation products play important roles in stress tolerance, plants respond to abiotic and abiotic stress by systematically accumulating higher levels primary and secondary metabolites for increasing their resistance. Glucosinolates play important role and have a relation with biotic and abiotic stress in *Brassica* plant family, as they can act as a signalling molecules and affect the physiology of plant.

Keywords: abiotic stresses, biotic stress, *Brassicaceae*, glucosinolates, stress tolerance, human health

1. Introduction

Plants are attached to the ground by the roots for water and nutrient supply and fixed at one place, so they have to face various abiotic stresses such as temperature, water availability, salt concentrations, etc. This is a common environmental stress which impacts productivity of the plants worldwide. Along with abiotic stresses, plants also face injuries by herbivory and insects that chew on plants. Once the plant faces mechanical damage by insects, a lot of changes occur in the plant [1]. In case of biotic stress, immediately after wounding by pest changes occur in the damaged area, which can be systematic (in the non-wounded areas) or local (in the tissues directly damaged). Once the injury occurs, metabolic changes, physiological changes and induction of gene expression started taking place [2]. The response to stress is not simple, rather involves different mechanisms and molecules. The strength and type of response depend upon plant species, developmental stage, nature of stress and environmental conditions. Plant develops, coordinates and combines all the defence mechanisms to cope with the challenges caused by the injuries [1, 3, 4].

The plant family *Brassicaceae* (or *Cruciferae*) includes some of the world's most economically important crops, especially members of the genera *Brassica* L. (Coles, mustards, oilseed rapes, turnips, etc.). These cultivated species include a large variety of leaf and root vegetables, oilseed crops which are consumed all over the world [5]. *Brassicaceae* vegetables are a good source of antioxidants because of their high phenolics and sulphur-containing glucosinolate content. These glucosinolates are responsible for characteristic flavour and odour of these vagetables. Myrosinase (thioglucoside glucohydrolase; EC3.2.1.1), an enzyme present in *Cruciferae*, is activated when cells are damaged by herbivory, cutting and chewing of insects. These Myrosinases hydrolyse glucosinolates into a variety of products, which generally include isothiocyanates, nitriles, epithionitriles and oxazoldine-2-thiones [6]. Glucosinolates and their degradation products play important roles in pathogen and insect interactions, stress tolerance as well as in human health. These compounds have chemoprotective properties against chemical carcinogens. Their role in prevention of cardiovascular disease is due to properties such as inhibiting the enzyme activation, modifying the steroid hormone metabolism and protecting against oxidative damage [7, 8]. Isothyocianate and indole products formed from glucosinolates hydrolysis by Myrosinase may regulate cancer cell development by regulating target enzymes, controlling apoptosis and blocking the cell cycle [9].

Various biotic and abiotic factors activate the defence system in plants, and these result in significant variation in the plant metabolome. These factors result in enhanced production of certain metabolites, e.g., amino acids, sugars, indoles, phenolics including glucosinolates. Glucosinolates production in stressed plants is seen to be enhanced, and they have a crucial role to play in mitigating the negative effect of stress. The concentration and type of glucosinolates and their hydrolysis in plants are affected by many factors, such as age, abiotic stress such as heavy metal, water availability, postharvest storage signalling molecules [10, 11]. These play an important role in abiotic stress. Their levels change in response to a variety of abiotic stress tolerances [12]. They also play an important role in plant-insect interactions. Their levels also change in response to herbivory in plants [13].

2. Glucosinolates and abiotic stress

Plants synthesise a variety of secondary metabolites that do not seem to have any essential role in plants, but presence of these compounds plays a very important role in plants and has many advantages, especially in defence. Glucosinolates, a class of secondary metabolites mainly found in Brassica family, are nitrogen and sulphur-containing compounds derived from amino acids such as methionine, alanine, valine or leucine (aliphatic); phenylalanine or tyrosine (aromatic) and tryptophan (indolic glucosinolates). These glucosinolates are usually increased in different biotic or abiotic stress conditions and play an important role in plant defence in these unfavourable conditions [14].

During growth, plants are exposed to various biotic (herbivory, fungal, bacterial and/or viral infection) and abiotic (metals, UV, temperature) stresses. It leads to upregulation of various gene and biochemical changes, which finally results in an enhancement of the synthesis of primary and secondary metabolites. Plants have acquired various stress tolerance mechanisms, involving physiological and biochemical changes that result in adaptive or morphological changes [15]. It was also found that plants respond to abiotic stress by accumulating higher or lower levels of Glucosinolates. The intensity and duration of the abiotic stress, as well as the developmental stage of the plant at the moment of the imposed stress, are important factors in the accumulation of each specific Glucosinolates [16, 17].

The mechanism by which the plant resists these stresses with help of metabolites is unclear as their regulation is a complex process because glucosinolates synthesis pathways are linked to various other pathways making it more complex to understand its direct role. There are studies in elucidating a relationship between the metabolites and stress, but there is a strong need to study underground mechanisms as well.

2.1 Salt stress

One of the major abiotic stresses affecting plant growth and development is salt stress [18]. There are various ways by which plants respond to this stress conditions either by osmotic adjustments or by accumulating secondary metabolites, such as Glucosinolates [19].

In one of the studies, Broccoli was grown in greenhouse condition and was treated with NaCl (0, 40, 80, 100 mM) to study the effect of Glucosinolates in salt-stressed Broccoli. Glucosinolates were determined at three different time points (1, 3, 6 days upon treatment), and it was found that the total aliphatic and indoles were lower at 40, 80 and 100 mM NaCl applications compared with control plants determined at three different time points [20]. A study was conducted with two Broccoli cultivars *cv Parthenon* and *cv. Naxos* grown in controlled conditions in Hoagland solutions with five salt treatments for 15 days. To avoid osmotic shock, 30 mM salt solution was added every hour until the final salt concentrations of 0, 30, 60 and 90 mM were reached. It was found that *cv Naxos* was more sensitive to salt as compared with *cv Parthenon*, and the indole glucosinolates were not affected by salinity in Parthenon while it decreased significantly in *cv Naxo*, whereas *Naxos cv* accumulated more aliphatic GSLs under salt stress than *Parthenon cv* suggesting that each broccoli cultivar adopted distinct strategy, which can be dependent on various factors [16].

Greater accumulation of proline and glucosinolates was found with increased salinity level in Canola or oilseed rape (*Brassica napus* L.), but it was dependent on the age and stage of plant growth—seedling, vegetative and reproductive. 131 *Brassica napus* genotypes were evaluated at these stages for glucosinolates with 0, 50, 100, 150, 200, 250, 300 and 350 mM NaCl concentration. It was concluded in this study that salt stress genotype should be selected at different stages as it is a developmental phenomenon for improvement of canola genotype productivity [21].

Six types of Brassica rocket genotypes used for salad belonging to the *Brassicaceae* family were tested at 0, 65 and 130 mM NaCl treatment for morphological parameter as well as different Glucosinolates concentrations. Morphological parameters were different for different genotypes, but it was observed that Glucosinolates content increased in moderate saline treatment, indicating that these rocket salads can be improved for good Glucosinolates content by irrigating them with salt water [22]. Increased level of Glucosinolates were shown to be increased in Radish, Kale and other *Brassicaceae* family as demonstrated by work done [23–25]. So it is clear from the above studies that at moderate salt stress, Glucosinolates content of the plant is increased and which in turn helps the plant in fighting stress tolerance along with other factors such as morphological other secondary metabolites and accumulating compatible solutes such as proline.

2.2 Water stress

Drought is unavailability of adequate amount of water to a plant, required for its proper growth and development. Drought in agriculture field can occur due to a variety of reasons majorly including low precipitation, inadequate ground water. When *Brassica* family is tested for Glucosinolates content in drought conditions, it was found out that it increases with drought in different varieties of *Brassica*. One such study was done on molecular mechanism in understanding the gene upregulated in response to drought conditions in Chinese cabbage, and it was found that glucosinolates content also increased with the drought stress indicating its possible role in drought [26, 27]. Even when table drought was given to *Arabidopsis* by PEG 6000, they showed an increase in glucosinolates concentration indicating their role in drought stress response, but this response was short-term response. When the drought stress increases, glucosinolates content decreases [28].

2.3 Heat stress

Abnormally high temperature during hot weather is again one kind of abiotic stress. Effect of high temperature depends on the duration and its intensity. Heat stress may cause various physical, physiological and molecular damage to the plants and in turn affects its productivity. In a study on *Arabidopsis* mutant TU8, (Glucosinolate-deficient and heat-sensitive mutant) [29], it was found from research that due to ineffective expression of heat shock protein (HSP), these mutants were unable to survive in high temperature and were glucosinolate-deficient too [30]. When overexpressing a protein phosphate gene in *Brassica rapa BrPP5.2* and thereby developing a transgenic line, it was observed that these plants develop a heat tolerance and high glucosinolates indicating role of glucosinolates in stress tolerance [31]. In other studies by different research groups, it was found that a variety of different glucosinolates content increased upon long-term and short-term heat shock in *Brassica* plants [32, 33]. As is a major part of the human diet and considering the health-benefiting glucosinolates in sprouts, this was increased by giving heat stress to developing sprouts. Heat shock at 60 degree C, in developing sprouts of cabbage, improves the isothiocyanates production and in turn the glucosinolates [34].

2.4 Heavy metal stress

Use and need of metals in plants are important as they act as cofactors in various enzymes such as Mg2+, Mn2+, Co2+, Zn2+, but with increase in industrialisation use of heavy metals such as Hg, Pb, Cd, Ni Ba Cr which are highly toxic even at a very low concentration. There are micronutrient essential metals such as Mn, Mo, Zn, Cu, Co which are required by plants in low amounts, but they also become toxic at high concentrations. Effect of cadmium toxicity was evaluated on cabbage and kale, and it was found that glucosinolates accumulate in stems as compared with other organs such as root or leaves, which indicate that glucosinolates can play role in heavy metal stress tolerance as plants have shown no metal toxicity physically [35]. In another study, *Brassica oleracea* plants were treated with solutions of ZnSO4 and CdSO4 by mixing them in soil for treatment of plants with the heavy metals Zn and Cd, and it was found that glucosinolates content increased as the concentration of Zn and Cu increased [36]. Increased concentration of cadmium was studied on Brassica napus, and it was found that glucosinolates concentration increased with increase in concentration of heavy metal [37]. Chinese cabbage when exposed to elevated Zn concentration was shown to overexpress the genes, APS reductase

(APR), associated with the biosynthesis of primary sulphur compounds. There was also an increase in aliphatic glucosinolates with an increase in gene expression of key biosynthetic enzymes and regulators such as CYP79B3, CYP83B1, MYB34 [38]. Brassica plants were also used for phytoremediation in heavy metal-contaminated soil, and it was also found as mentioned in above studies that glucosinolates concentration increases with increase in concentration of heavy metal. It can also be stated that accumulation of glucosinolates can be one of the strategies for heavy metal tolerance in *Brassica* plants.

2.5 UV stress

Plant growth and development are mainly determined by sunlight. Sunlight is used in germination, development, photomorphogenesis and various other important functions in plants. But as sunlight is composed of UV-A, UV-B and UV-C, these radiations cause stress in plants. As a result of these radiation stresses, DNA damage, generation of reactive oxygen species (ROS) and many other cellular changes take place in plants. Glucosinolates also get affected by stress caused by UV light, as proven with the studies done on broccoli sprouts which were exposed to various low- and high-intensity UVA and UVB lamps for 120 min which resulted in high levels of glucosinolates in broccoli sprouts [39]. Even the combined effects of Methyl Jasmonate (MJ), UVA or UVB lights enhance the glucosinolates in broccoli sprouts [40]. As it is known from the studies that glucosinolates levels are upregulated during the UV stress [41], and it was also associated with increased biotic stress tolerance too. Low-to-medium UV-B doses were applied to sprouts of broccoli, and it was found that there was an increase in glucosinolate levels of 4-methylsulfinyl butyl GS and 4-methoxy-indol-3-ylmethyl GS. These glucosinolates in turn affect the genes related to jasmonic acid and salicylic acid signalling pathway, which ultimately leads to overexpression of bacterial and fungal pathogen-responsive genes [42].

3. Glucosinolates and biotic stress

Biotic stress in plants is caused by any living organisms, such as insects, pests, viruses, bacteria, fungi, nematodes, etc. Microorganisms damage plants by causing symptoms such as plant wilt, leaf spots, root rot, etc. Insects can cause physical damage to plants as they either chew or suck on the plants, including the leaves, stem, bark and flowers. They can also act as a vector of viruses and bacteria from infected plants to healthy plants. As plants do not have any immune systems to deal with this kind of stress, they respond to this stress by developing many changes at morphological, chemical and protein levels. The main aim of these changes is to protect the plants from foreign invaders. One such chemical is glucosinolates which get elevated in response to biotic stress for the purpose of defence in plants.

Chemically, glucosinolates are composed of thiohydroximate-O-sulfonate group linked to glucose and an alkyl, aralkyl or indolyl side chain (R). More than 200 glucosinolates have been identified in Brassica crops, and they are characterised mainly by the variable R group, which can be aromatic, indolic or aliphatic derivatives of the amino acid precursors methionine, tryptophan and phenylalanine, respectively [43, 44].

Glucosinolates play a variety of biological functions in plants [45] and get converted into various degradation products (isothiocyanates, thiocyanates, indoles, etc.), when vegetables containing them are cut or chewed because during this process, they come in contact with the enzyme myrosinase which hydrolyses them, as discussed earlier in this chapter. A sudden outbreak of these insecticidal compounds in field is called as **'mustard oil bomb'** [46]. These degradation products have both harmful and beneficial roles in plants as well as in human health. These compounds protect plants against various pathogens and weed but are also responsible for the bitter flavour of the crucifers because of the presence of some glucosinolates such as sinigrin and progoitrin [47]. Glucosinolates can act as potent anticarcinogens too [48, 49].

Studies have determined the effect of insect pest and glucosinolates levels in *Brassica* [50, 51]. These kinds of studies were done directly on the glucosinolates level in plant tissue and with the study of expression of gene family which is associated with the glucosinolates and its associated enzymes. To study the biotic stress, signalling molecules such as plant hormones, salicylic acid (SA), jasmonic acid (JA) and ethylene play very important roles. In case of insects and pest, mechanical injury is mimicked with the help of damage through forceps and needles to plants. It has been proved that plant signalling molecules such as methyl jasmonate, salicylic acid and glucose as well as mechanical injury affect glucosinolates levels in *Arabidopsis* and *Brassica* species [52–55]. Therefore, these signalling molecules are used for the study of biotic stress in plant in laboratories.

Several reports exist on the toxicity of Glucosinolates hydrolysis products to bacteria and fungi. The activity of Glucosinolates and their degraded products against various strains of microorganisms has been documented by many investigators being present in the leaves of *Brassica* spp. at concentrations that can prevent the development of pathogens [56].

3.1 Glucosinolates and pest

When generalist herbivores feed on Brassica, degradation product of Glucosinolates in them, the isothiocyanates have toxic effects and function as repellents. However, specialist herbivores use Glucosinolates as attractants, and they can survive on plants containing them. For example, *Brevicoryne brassicae* aphids, flea beetles, *Psylliodes chrysocephala* and the Lepidopteran pest, *Pieris rapae* are some examples of specialist herbivores [57].

It was also found that plants respond to herbivore or insect damage by systematically accumulating higher levels of glucosinolates and thus presumably increasing their resistance. Usually it is the indole glucosinolates which become induced [58, 59]. Sometimes volatiles produced by Glucosinolates provide indirect protection to plants by attracting natural enemies of herbivores such as parasitoids [60].

Lepidopteran that feeds almost exclusively on *Brassicaceae* plants is *Plutella xylostella* (Diamond Moth Back). A study done by [61] compares two lepidopteran larvae, one specialist and one generalist, namely *Mamestra brassicae* (Lepidoptera, Noctuidae) and the specialist *Pieris rapae* (Lepidoptera, Pieridae) respectively with respect to six different genotypes of *Brassica oleracea.* They found that two aliphatic (sinigrin and glucoiberin) and one indole (glucobrassicin) glucosinolates affect the larvae growth which in turn is affected by the age of the *Brassica oleracea* plants. Mature plants produce more Glucosinolates and are protected from the pest as compared with the young plants which is determined by larval development time, body weight, mortality and feeding rate of two larvae selected for study.

Genetic engineering of glucosinolates biosynthesis pathway for the production of glucosinolates in non-Glucosinolates producing plants was successfully employed by [62]. Genetic transformation of *Nicotiana tabacum* (tobacco) for the benzyl glucosinolate pathway was done from *A. thaliana* which results in successful production of benzyl glucosinolate in tobacco, which attracted the female pest *Plutella xylostella* (diamondback moth). Larvae hatched that could not survive on tobacco plants demonstrate that this strategy can be used as the trap crop for many important crucifer crops.

It was found that *Brassica* plants were used for the purpose of Biofumigation in fields which include suppression of pests and pathogens through release of volatile substances from degradation of biomass into the soil. *Brassica* plant contains higher amount of Glucosinolates and its breakdown product which are released after tissue damage, such as isothiocyanates which are toxic, and also they have the fumigant property which is comparable with metham-sodium [63–65].

4. Conclusion

Abiotic stress such as heat, drought and high salt and biotic stress including various attacks on plant from insects, pest or other bacterial viral disease can affect the plant in terms of physical, physiological and many other biochemical and molecular aspects, which can retard its growth and productivity.

Brassicaceae family includes important crop due its medicinal and nutritional properties. This family has plants which are valuable source of oil and food containing medicinal values including anti-cinereous property for humans. These crop plants like other crop plant face abiotic and biotic stress both, and they use various mechanisms to overcome these stresses. Glucosinolates, which are responsible for the pungent characteristic smell of *Brassica*, are responsible for various biotic and abiotic stress tolerances in Brassica. They have particular pathway from which they are synthesised and regulated by various factors. MYB group of transcriptional factors are involved in glucosinolates synthesised pathway [66]. Environment, signalling molecules, herbivores and pathogens also regulate the glucosinolates biosynthesis which can help in their stress tolerance by upregulating different transcription factors and genes. In case of biotic stress, glucosinolates in the form of 'mustard oil bomb' can provide the defence against generalist and specialist herbivore. In spite of so many studies and research done, there are many gaps yet to be identified and many conclusions need to be explored in depth. Role of glucosinolates can be much explored as nutrition, medicinal and biocontrol agents in field. Glucosinolates also play important role in stress tolerance in Brassica, understanding their deeper role in abiotic and biotic stress can help in increasing their productivity worldwide.

Author details

Parul Chowdhury
Dr. B. Lal Institute of Biotechnology, Affiliated to University of Rajasthan, Jaipur, Rajasthan, India

*Address all correspondence to: parul.bibt@gmail.com

References

[1] León J, Rojo E, Sánchez-Serrano JJ. Wound signalling in plants. Journal of Experimental Botany. 2001;**52**(354):1-9

[2] Nguyen VP, Stewart J, Lopez M, Ioannou I, Allais F. Glucosinolates: Natural occurrence, biosynthesis, accessibility, isolation, structures, and biological activities. Molecules. 2020;**25**(19):4537

[3] Gull A, Lone AA, Wani NU. Biotic and abiotic stresses in plants. In: Abiotic and Biotic Stress in Plants. Chapter 6. UK: Intech open; 2019. pp. 1-9

[4] He Q, Bertness MD, Altieri AH. Global shifts towards positive species interactions with increasing environmental stress. Ecology Letters. 2013;**16**(5):695-706

[5] Rakow G. Species origin and economic importance of brassica. In: Brassica. Berlin, Heidelberg: Springer; 2004. pp. 3-11

[6] Jahangir M, Abdel-Farid IB, Kim HK, Choi YH, Verpoorte R. Healthy and unhealthy plants: The effect of stress on the metabolism of Brassicaceae. Environmental and Experimental Botany. 2009;**67**(1):23-33

[7] Jongen WM. Glucosinolates in brassica: Occurrence and significance as cancer-modulating agents. Proceedings of the Nutrition Society. 1996; **55**(1B):433-446

[8] Das S, Tyagi AK, Kaur H. Cancer modulation by glucosinolates: A review. Current Science. 2000;**79**:1665-1671

[9] Cartea ME, Velasco P. Glucosinolates in brassica foods: Bioavailability in food and significance for human health. Phytochemistry Reviews. 2008;**7**(2): 213-229

[10] Kang JY, Ibrahim KE, Juvik JA, Kim DH, Kang WJ. Genetic and environmental variation of glucosinolate content in Chinese cabbage. HortScience. 2006; **41**(6):1382-1385

[11] Sun B, Yan H, Liu N, Wei J, Wang Q. Effect of 1-MCP treatment on postharvest quality characters, antioxidants and glucosinolates of Chinese kale. Food Chemistry. 2012;**131**(2):519-526

[12] del Carmen M-BM, Moreno DA, Carvajal M. The physiological importance of glucosinolates on plant response to abiotic stress in brassica. International journal of molecular sciences. 2013;**14**(6):11607-11625

[13] Halkier BA, Gershenzon J. Biology and biochemistry of glucosinolates. Annual Review of Plant Biology. 2006;**57**:303-333

[14] Arbona V, Manzi M, Ollas CD, Gómez-Cadenas A. Metabolomics as a tool to investigate abiotic stress tolerance in plants. International journal of molecular sciences. 2013; **14**(3):4885-4911

[15] Hasanuzzaman M, Nahar K, Alam M, Roychowdhury R, Fujita M. Physiological, biochemical, and molecular mechanisms of heat stress tolerance in plants. International journal of molecular sciences. 2013;**14**(5): 9643-9684

[16] Zaghdoud C, Alcaraz-Lopez C, Mota-Cadenas C, Martínez-Ballesta MD, Moreno DA, Ferchichi A, et al. Differential responses of two broccoli (Brassica oleracea L. varItalica) cultivars to salinity and nutritional quality improvement. The Scientific World Journal. 2012;**2012**:1-12

[17] Cocetta G, Mishra S, Raffaelli A, Ferrante A. Effect of heat root stress and high salinity on glucosinolates

metabolism in wild rocket. Journal of Plant Physiology. 2018;**231**:261-270

[18] Zhu JK. Plant salt tolerance. Trends in plant science. 2001;**6**(2):66-71

[19] Velasco P, Cartea ME, González C, Vilar M, Ordás A. Factors affecting the glucosinolate content of kale (brassica oleraceaacephala group). Journal of agricultural and food chemistry. 2007;**55**(3):955-962

[20] Sarikamiş G, Cakir A. Influence of salinity on aliphatic and indole glucosinolates in broccoli (Brassica oleracea var. italica). Applied Ecology and Environmental Research. 2017;**15**(3):1781-1788

[21] Gyawali S, Parkin IA, Steppuhn H, Buchwaldt M, Adhikari B, Wood R, et al. Seedling, early vegetative, and adult plant growth of oilseed rapes (Brassica napus L.) under saline stress. Canadian Journal of Plant Science. 2019; **99**(6):927-941

[22] Petretto GL, Urgeghe PP, Massa D, Melito S. Effect of salinity (NaCl) on plant growth, nutrient content, and glucosinolate hydrolysis products trends in rocket genotypes. Plant Physiology and Biochemistry. 2019;**141**:30-39

[23] Yuan G, Wang X, Guo R, Wang Q. Effect of salt stress on phenolic compounds, glucosinolates, myrosinase and antioxidant activity in radish sprouts. Food Chemistry. 2010; **121**(4):1014-1019

[24] Guo RF, Yuan GF, Wang QM. Effect of NaCl treatments on glucosinolate metabolism in broccoli sprouts. Journal of Zhejiang University Science B. 2013;**14**(2):124-131

[25] Maina S, Ryu DH, Cho JY, Jung DS, Park JE, Nho CW, et al. Exposure to salinity and light spectra regulates glucosinolates, phenolics, and antioxidant capacity of Brassica carinata L. microgreens. Antioxidants. 2021;**10**(8):1183

[26] Eom SH, Baek SA, Kim JK, Hyun TK. Transcriptome analysis in Chinese cabbage (Brassica rapa ssp. pekinensis) provides the role of glucosinolate metabolism in response to drought stress. Molecules. 2018; **23**(5):1186

[27] Shawon RA, Kang BS, Lee SG, Kim SK, Lee HJ, Katrich E, et al. Influence of drought stress on bioactive compounds, antioxidant enzymes and glucosinolate contents of Chinese cabbage (Brassica rapa). Food chemistry. 2020;**308**:125657

[28] Sun J, Yang J, Yu T. Effects of drought stress simulated by PEG on glucosinolates content in Arabidopsis thaliana. Agricultural Science & Technology-Hunan. 2010;**11**(3):20-24

[29] Haughn GW, Davin L, Giblin M, Underhill EW. Biochemical genetics of plant secondary metabolites in Arabidopsis thaliana: The glucosinolates. Plant Physiology. 1991;**97**(1):217-226

[30] Ludwig-Muller J, Krishna P, Forreiter C. A glucosinolate mutant of Arabidopsis is thermosensitive and defective in cytosolic Hsp90 expression after heat stress. Plant Physiology. 2000;**123**(3):949-958

[31] Muthusamy M, Kim J, Kim S, Park S, Lee S. BrPP5. 2 overexpression confers heat shock tolerance in transgenic Brassica rapa through inherent chaperone activity, induced Glucosinolate biosynthesis, and differential regulation of abiotic stress response genes. International Journal of Molecular Sciences. 2021;**22**(12):6437

[32] Rao SQ, Chen XQ, Wang KH, Zhu ZJ, Yang J, Zhu B. Effect of short-term high temperature on the accumulation of glucosinolates in

Brassica rapa. Plant Physiology and Biochemistry. 2021;**161**:222-233

[33] Kask K, Kännaste A, Talts E, Copolovici L, Niinemets Ü. How specialized volatiles respond to chronic and short-term physiological and shock heat stress in Brassica nigra. Plant, cell & environment. 2016;**39**(9):2027-2042

[34] Yang R, Guo L, Wang J, Wang Z, Gu Z. Heat shock enhances isothiocyanate formation and antioxidant capacity of cabbage sprouts. Journal of Food Processing and Preservation. 2017;**41**(4):e13034

[35] Jakovljević T, Cvjetko M, Sedak M, Đokić M, Bilandžić N, Vorkapić-Furač J, et al. Balance of glucosinolates content under Cd stress in two brassica species. Plant Physiology and Biochemistry. 2013;**63**:99-106

[36] Kusznierewicz B, Bączek-Kwinta R, Bartoszek A, Piekarska A, Huk A, Manikowska A, et al. The dose-dependent influence of zinc and cadmium contamination of soil on their uptake and glucosinolate content in white cabbage (Brassica oleracea var. capitata f. alba). Environmental Toxicology and Chemistry. 2012;**31**(11):2482-2489

[37] Durenne B, Druart P, Blondel A, Fauconnier ML. How cadmium affects the fitness and the glucosinolate content of oilseed rape plantlets. Environmental and Experimental Botany. 2018;**155**:185-194

[38] Aghajanzadeh TA, Prajapati DH, Burow M. Differential partitioning of thiols and glucosinolates between shoot and root in Chinese cabbage upon excess zinc exposure. Journal of Plant Physiology. 2020;**244**:153088

[39] Moreira-Rodríguez M, Nair V, Benavides J, Cisneros-Zevallos L, Jacobo-Velázquez DA. UVA, UVB light doses and harvesting time differentially tailor glucosinolate and phenolic profiles in broccoli sprouts. Molecules. 2017;**22**(7):1065

[40] Moreira-Rodríguez M, Nair V, Benavides J, Cisneros-Zevallos L, Jacobo-Velázquez DA. UVA, UVB light, and methyl jasmonate, alone or combined, redirect the biosynthesis of glucosinolates, phenolics, carotenoids, and chlorophylls in broccoli sprouts. International Journal of Molecular Sciences. 2017;**18**(11):2330

[41] Rechner O, Neugart S, Schreiner M, Wu S, Poehling HM. Can narrow-bandwidth light from UV-A to green alter secondary plant metabolism and increase brassica plant defenses against aphids? PLoS One. 2017;**12**(11):e0188522

[42] Mewis I, Schreiner M, Nguyen CN, Krumbein A, Ulrichs C, Lohse M, et al. UV-B irradiation changes specifically the secondary metabolite profile in broccoli sprouts: Induced signaling overlaps with defense response to biotic stressors. Plant and Cell Physiology. 2012;**53**(9):1546-1560

[43] Fahey JW, Zalcmann AT, Talalay P. The chemical diversity and distribution of glucosinolates and isothiocyanates among plants. Phytochemistry. 2001;**56**(1):5-1

[44] Latté KP, Appel KE, Lampen A. Health benefits and possible risks of broccoli–an overview. Food and Chemical Toxicology. 2011;**49**(12):3287-3309

[45] Bhandari SR, Kwak JH. Chemical composition and antioxidant activity in different tissues of brassica vegetables. Molecules. 2015;**20**(1):1228-1243

[46] Müller C, Brakefield PM. Analysis of a chemical defense in sawfly larvae: Easy bleeding targets predatory wasps in late summer. Journal of chemical ecology. 2003;**29**(12):2683-2694

[47] Tripathi MK, Mishra AS. Glucosinolates in animal nutrition: A review. Animal feed science and technology. 2007;**132**(1-2):1-27

[48] Higdon JV, Delage B, Williams DE, Dashwood RH. Cruciferous vegetables and human cancer risk: Epidemiologic evidence and mechanistic basis. Pharmacological Research. 2007; **55**(3):224-236

[49] Traka M, Mithen R. Glucosinolates, isothiocyanates and human health. Phytochemistry Reviews. 2009; **8**(1):269-282

[50] Siemens DH, Mitchell-Olds T. Glucosinolates and herbivory by specialists (Coleoptera: Chrysomelidae, Lepidoptera: Plutellidae): Consequences of concentration and induced resistance. Environmental Entomology. 1996; **25**(6):1344-1353

[51] Mewis I, Appel HM, Hom A, Raina R, Schultz JC. Major signaling pathways modulate Arabidopsis glucosinolate accumulation and response to both phloem-feeding and chewing insects. Plant Physiology. 2005;**138**(2):1149-1162

[52] Taipalensuu J, Falk A, Rask L. A wound-and methyl jasmonate-inducible transcript coding for a myrosinase-associated protein with similarities to an early nodulin. Plant Physiology. 1996;**110**(2):483-491

[53] Bodnaryk RP. Effects of wounding on glucosinolates in the cotyledons of oilseed rape and mustard. Phytochemistry. 1992;**31**(8):2671-2677

[54] Bartlet E, Kiddle G, Williams I, Wallsgrove R. Wound-induced increases in the glucosinolate content of oilseed rape and their effect on subsequent herbivory by a crucifer specialist. In: Proceedings of the 10th International Symposium on Insect-Plant Relationships. Dordrecht: Springer; 1999. pp. 163-167

[55] Pontoppidan B, Hopkins R, Rask L, Meijer J. Differential wound induction of the myrosinase system in oilseed rape (Brassica napus): Contrasting insect damage with mechanical damage. Plant Science. 2005;**168**(3):715-722

[56] Brader G, Tas E, Palva ET. Jasmonate-dependent induction of indole glucosinolates in Arabidopsis by culture filtrates of the nonspecific pathogen Erwiniacarotovora. Plant Physiology. 2001;**126**(2):849-860

[57] Bruce TJ. Glucosinolates in oilseed rape: Secondary metabolites that influence interactions with herbivores and their natural enemies. Annals of Applied Biology. 2014;**164**(3):348-353

[58] Giamoustaris A, Mithen R. The effect of modifying the glucosinolate content of leaves of oilseed rape (Brassica napus ssp. oleifera) on its interaction with specialist and generalist pests. Annals of Applied Biology. 1995;**126**(2):347-363

[59] Martin N, Müller C. Induction of plant responses by a sequestering insect: Relationship of glucosinolate concentration and myrosinase activity. Basic and Applied Ecology. 2007;**8**(1):13-25

[60] De Bortoli SA, Polanczyk RA, Vacari AM, De Bortoli CP, Duarte RT. Plutellaxylostella (Linnaeus, 1758) (Lepidoptera: Plutellidae): Tactics for integrated pest management in Brassicaceae. In: Weed and Pest Control-Conventional and New Challenges. Rijeka: InTech; 2013. pp. 31-51

[61] Santolamazza-Carbone S, Sotelo T, Velasco P, Cartea ME. Antibiotic properties of the glucosinolates of Brassica oleracea var. acephala similarly affect generalist and specialist larvae of two lepidopteran pests. Journal of Pest Science. 2016;**89**(1):195-206

[62] Møldrup ME, Geu-Flores F, de Vos M, Olsen CE, Sun J, Jander G, et al.

Engineering of benzylglucosinolate in tobacco provides proof-of-concept for dead-end trap crops genetically modified to attract Plutellaxylostella (diamondback moth). Plant biotechnology journal. 2012; **10**(4):435-442

[63] Kirkegaard JA, Sarwar M. Biofumigation potential of brassicas. Plant and Soil. 1998;**201**(1):71-89

[64] Smolinska U, Morra MJ, Knudsen GR, James RL. Isothiocyanates produced by Brassicaceae species as inhibitors of fusarium oxysporum. Plant disease. 2003;**87**(4):407-412

[65] Prasad P, Kumar J. Management of Fusarium wilt of chickpea using brassicas as biofumigants. Legume Research-An International Journal. 2017;**40**(1):178-182

[66] Hirai MY, Sugiyama K, Sawada Y, Tohge T, Obayashi T, Suzuki A, et al. Omics-based identification of Arabidopsis Myb transcription factors regulating aliphatic glucosinolate biosynthesis. Proceedings of the National Academy of Sciences. 2007;**104**(15):6478-6483

Chapter 9

Plant Proteome in Response to Abiotic Stress

Fatemeh Habibpourmehraban

Abstract

Due to their sessile nature, plants have to confront the stresses and develop potent adaptive tactics to survive and thrive or tolerate their adverse effects. Abiotic stresses, pose a severe threat and multiple morphologies, biochemistry, and physiology procedures to agriculture and the ecosystem. On the other hand, reductions in crop yields brought about by abiotic stress are expected to increase as climate change restricts the worldwide utilization of arable lands and indirectly affects crop productivity. Therefore, understanding how plants perceive stress signals and adapt to unfavorable environmental conditions is crucial for future global food safety and security. In this chapter, we summarize the latest findings of the effects of abiotic stresses on molecular changes in plant organisms, cells, and tissues, focusing on the stress-specific sensing biomolecules and mechanisms at the proteome level.

Keywords: abiotic stress, plant, drought, salt, cold, high temperature, proteomics

1. Introduction

Food could be a reason for the third world war as severe land competition areas participate in agriculture and habitation. Moreover, food production depends not only on land provided but also on the institutionalization of sustainable crop production related to food safety and security.

It is predicted that by 2050, the world's population will grow to nearly 10 billion people, and the main issue will be feeding them sustainably. To meet this demand, it is necessary to produce about 50% more food in less than 40 years. And the parameters affecting food production like climate change, scarcity of natural resources, and food-wasting should be considered carefully.

On the other hand, approaching sustainable food availability in agriculture not only depends on a balance between food production sustainability, food security, and food safety, but also to spread food worldwide justly [1, 2].

Hunger and undernutrition are two main consequences of food insecurity. So, finding the solutions to make long-term sustainable agriculture possible concerning its restricting environmental factors is a severe challenge for future studies. As one of the significant environmental factors, climate change is expected to be the most unfavorable challenge for sustainable crop production with consistent and adverse effects on food security in many countries [3, 4]. Climate change could be explained by higher precipitation variability, increasing droughts, or floods accompanied by temperature fluctuation. Findings illustrated that about 90% of arable lands are prone to one or more of the above stresses [5], which cause up to 70% yield losses in major food crops [6].

IntechOpen

Agriculture and climate have mutual interaction by affecting each other both positively and negatively. Although, the agriculture industry impacts climate in two main contributions, including leading to approximately 25% of global anthropogenic emissions [7] and about 70% of global water withdrawals [8]. The most serious challenge to reach sustainable agriculture would be environmental stresses with emphasizing abiotic stresses.

Regarding climate impression on-farm activities, climate change could affect crop productivity through direct and indirect pathways. Changes in temperature, water availability, and greater variation in weather conditions are significant direct impact factors [9]. For example, temperature increases cause faster plant growing, shorter cropping seasons, and lower yields subsequently. Moreover, pathways like variation in pests, pathogens, and pollinators could be named indirect effects of environment change on crop yield [10]. For instance, a meta-analysis of 1090 studies on the yield of principal crops subjected to unfavorable environmental conditions confirmed that yield reduction could remarkably happen in long-term agriculture [3].

Environmental alterations' direct and indirect effects could be managed by developing adaptation mechanisms like using new plant production methods or breeding new plant genotypes resistant/low sensitive to environmental pressure [11]. Identifying and quantifying the role of abiotic stresses on the future of plant products may be detected by using technologies to study plants deeply and finding tolerant related genes, biomarkers, or metabolites to work. Proteomics is a reliable and accurate technique for investigating plants' responses to various stresses and detecting mechanisms specific to each genotype, stress, or combination of them.

In this chapter, we want to focus on the effects of climate change on agriculture, specifically on the opposing side therefore, various type of abiotic environmental factors limiting plant production at proteome level will be explained with detail and proteomics technology that helps study plants' proteome profile under stress conditions.

2. Environmental stressors

The environment could positively and negatively affect populations, organisms, and ecosystems; that negative impact is named stressors with variation in intensities. Exposure is the interaction across each organism with an environmental stressor that is specific to time and location. Exposure is defined as short term and long term with intensity variation. Suppose exposure leads to different types of internal or external changes, named as the response. Various categories of environmental stressors have been recognized, including climatic stress, biological stressors, biological pollution, physical stress, wildfire, chemical pollution, thermal pollution, and radiation stress [12]. Climatic stress is the primary factor affecting crop production worldwide by expressing drought, flood, heat, and cold.

In contrast with human beings, farmers cannot control environmental stresses and have to accept them as nature, so studies have been focused on finding mechanisms in plants to keep them productive in average conditions and under unfavorable situations.

3. Environmental stresses on plants

One of the most critical challenges in plant production is the competition of specifying lands for farms or growing population habitations. So, it is vital to produce more in less area considering climate change and anthropogenic activities consequences [13]. Alteration in natural elements like temperature increasing by 3–5°C in the next 50–100 years is proof of extending drought tress, besides human activities such as increasing consumption of chemical fertilizers, pesticides, and improper usage of groundwater resources resulting in salt stress. Therefore, increasing environmental stresses is inescapable from plant productivity in the future [14]. It should be noticed that stress impact not only depends on environmental conditions but also on plant genotype. Therefore, plant reaction to stress is a specific interaction of genotype × environment that could vary depending on these two parameters [15].

In another way, animals counter to negative pressures by escaping and moving, but for immobile plants, stress is the alarm of the typical environment modifying to uncarvable status. Generally, two types of environmental stress, including biotic stress and abiotic stress, are categorized that biotic stress defines plant condition after subjecting to a biological invader. In contrast, nonliving environmental factors are imposed on plants as abiotic stresses [16]. Biotic and abiotic stresses are essential to be studied in detail as they are the main reasons for plant product loss globally [17]. It needs to be considered that the source of environmental pressure affects the synchronism, diversification, or even the extinction of plants, inevitably related to agriculture development [18].

3.1 Biotic stress on plants

Biotic stress results from competition for nutrition absorbing between plants and various aggressive range of pests and pathogens, including viruses, bacteria, fungi, nematodes, herbivorous insects, arachnids, and weeds [19, 20]. The level of plant tolerance depends on its balance in responding to biotic stress [21]. Plant starts to activate individual and combined mechanisms in different morphological, physiological, biochemical, and molecular levels, and interaction across these functions expresses plant sensing of stress [22–24].

3.2 Abiotic stress on plants

One of the significant climatic changes in the next 50–100 years is surface temperature increasing by 3–5°C that in combination with an increasing trend of drought, flood, and heatwaves, will be expected to influence crop productivity negatively and food safety [25–27]. For instance, drought and heat stress substantially affect seed yields by reducing seed size and number, consequently loss in trait '100 seed weight' and seed quality [28].

Environmental abiotic stressors include drought (water stress), excessive watering (waterlogging) and submergence, extreme temperatures [high and low (chilling, cold, freezing)], salinity due to excessive Na^+, deficient or over essential nutrients, chemical factors (heavy metals and pH), extreme levels of light (high and low), radiation (UV-B and UV-A), gaseous pollutants (ozone, sulfur dioxide), mechanical factors, and other less frequently occurring stressors trigger plants negatively by crop quantity and quality losing [5, 20, 29–34]. Importantly, abiotic stresses cause dramatic detriment in various species and some to 50% yield limitation [32, 33]. It should be noted that about 90% of farmlands are exposed to at least one of the above abiotic stresses [5].

3.2.1 Drought stress

Agriculture is accounting for 70% of total freshwater worldwide usage on average. Therefore, human water consumption competes with agriculture water demand, so it impresses the water availability for commercial plant production with the effect of drought stress [34]. The universal water deficiency directly limits plant production while earth temperature increasing with the trend of warming around 0.8°C over the past 100 years has an indirect effect. For example, global barley production reduces by 10% with each 1°C temperature increase [35, 36]. One of the approaches to accompany plants with growing drought stress could be water use efficiency (WUE). For instance, drought stress remarkably impacts nearly 23 million hectares of rain-fed rice-growing area in Southeast Asia [37]. By developing WUE, plants will use water more efficiently, suitable for drought stress tolerance [38]. Some of the primary effects of drought stress on plants are reducing plant growth rate, photosynthetic function, CO_2 concentration, and molecular metabolism [38].

3.2.2 Salt stress

Soil salinity is not only caused by environmental activities or factors like environmental pollution, especially in arid and semiarid arable lands but also could be related to drought stress by a deficiency in water resources. Salinity directly limits crop productivity, food safety, and agriculture sustainability by gradually cultivating salinized lands [39, 40]. The severity of salinization could be sensed better by finding that around 970 million hectares, 8% of arable lands, are impacted by a high level of salinity stress [41].

Specific effects of salt stress on plants could be explained physiologically by reducing seedling germination percentage, shoot and root length, fresh and dry weight, and necrotic leaf tissue morphologically and K^+ and Ca^{2+} level deficiency continued with osmotic and oxidative stresses at the molecular level, especially in plant leaves [42–45].

3.2.3 Temperature stress

Temperature stress is a geographical dependent variable that defines high-temperature stress and low-temperature stress. Suppose the climate condition declines to less than 15°C, known as chilling or cold stress, while by greater decreasing to less than <0°C, freezing stress happens. Generally speaking, low-temperature stress is detrimental stress detracting plant growth and yield by affecting germination, seedlings growing, the color of leaves, and tillering continued in declining pollen sterility [46, 47].

Against low temperature, increasing temperature to a higher level than plants tolerance modifies plant growth and productivity negatively. High temperature or heat stress disturbs a plant's average growth morphologically, physiologically, and in molecular aspects like protein degradation or modification, instability of enzymes, nucleic acids, biomembranes, and cytoskeletal structures [48].

3.2.3.1 Cold stress

Cold and freezing stresses affect plant production significantly by decreasing production or even plant death. The plant generally adapts to such severe stress conditions to survive; however, many plants cannot tolerate it [14]. Cold stress categorizes plants, with different resistance levels, into three groups: delicate chili plants that are highly sensitive to a low temperature just lower than 15°C and damage seriously. The second category, chilling resistant plants, which are medium tolerant to cold stress and are temperature around 0°C, causes stress. Nevertheless, the last plant group is frost-resistant plants that acclimate to shallow temperatures even by ice formation. Cold stress resulted in membrane instability, ion exchange disturbance, and electrolyte leakage [49].

3.2.3.2 Heat stress

Temperature increasing is a severe concern of future plant productivity. Heat stress impacts plant productivity variously in morphological, biochemical, and molecular levels. Plant growth and development decrease, seed germination decline beside photosynthetic lower efficiency all together trigger yield loss as a consequence of heat stress [14, 50]. Another effect of high temperature is accelerating plant growth, especially during the vegetative stage, to mature faster by fruit or seed production. Notably, heat stress could signal drought stress by increasing transpiration and, finally, water evaporation [51].

3.2.4 Heavy metal/metalloids

Soil plays a vital role in plant growth by providing nutrients; however, the amount of soil solution determines whether to call them nutrition. Heavy metal toxicity results from the high concentration of metalloids contaminate the soil with the possibility of absorbance into plant tissues and frustrating plant normal life cycle [52]. Unlikely, metal toxicity is predicted to cause mutagenic impacts on crops due to improper and sewage wastewater irrigation methods, redundant adding of chemical fertilizers to the soil, and rapid industrialization [14, 53]. Two main metal categories are recognized in soil that may lead to mental stress, namely vital micronutrients for average plant growth that become toxic if accumulating in excess in soil solution (Fe, Mg, Mo, Zn, Mn, Cu, and Ni) and non-essential elements with unknown physiological and biological function; however can damage the plant by accumulating in the soil even in modicum amounts (Ag, Cr, Cd, Co, As, Sb, Pb, Se, and Hg) [54–56]. These vital elements are crucial for enzyme and protein structure in plants, but their excessive presence is not helpful and causes abiotic stress in plants [57].

3.3 Effect of abiotic stress on plants

Plant adaptation in response to abiotic stresses includes a matrix and interaction of various morphological, biochemical, and molecular mechanisms. Morphological alterations are the visible symptoms, representing plant unfavorable condition after stress subjection. However, the number of unique responses happen in the plant adaptation process, some of the common effects like wilting due to water flow decrement, reduction in photosynthetic ability with the result of tiller number decreasing, reducing leaf growth, and increasing root length [58, 59].

Many biochemical mechanisms inducing adaption to stress in plants are regulated by increasing phytohormones levels [60]. Abscisic acid (ABA), jasmonic acid (JA), and ethylene are the main hormones leading to secondary stresses in plants such as osmotic and oxidative stresses [33, 61, 62]. Oxidative stress results from reactive oxygen species production, which generally happens in response to extreme temperature treatment. In contrast, salt and drought stresses lead to an osmotic imbalance in cells and cause osmotic stress [63]. By concomitant osmotic and oxidative stress under abiotic stresses, plant changes biochemically with some common responses like stomatal closure, reducing photosynthesis-related variables like gas exchange factors, declining photosynthesis, and increasing reactive oxygen scavenging activity [59]. Notably, the photosynthesis mechanism is a crucial physiologic parameter influence on yield output, and increasing in related variables helps plant adaptation experiencing unfavorable conditions [64, 65]. Positive induction of phytohormones in stress-subjected plants signals overexpression of the genes related to stress tolerance [66].

A comprehensive understanding of stress responses in plants needs to study the whole network interaction happening in plants involving individual or shared responses. Molecular mechanisms including proteome, transcriptome, genome, and metabolome modification analysis help find genes responsible for plant tolerance under abiotic stress [67, 68]. Genes encoding proteases, chaperonins, enzymes of sugar, proline, ion, and water channel proteins, enzymes contribute to oxidative stress, transcription factors (TFs), and protein kinases help to protect the plant against abiotic stress by overexpression [69]. Moreover, phytohormone signaling regulates some genes like ABA-dependent factor expression after exposure to abiotic stress. Basic leucine zipper (bZIP) TFs are an example of this category, leading tolerance-related responses like stomatal closure and expression of dehydration tolerance genes. However, there is another category of ABA-independent mechanisms [70].

3.4 Plant proteome profile alteration under abiotic stress

3.4.1 Proteomics technology

As the proteome is dynamic in plants both in control and stress conditions, its profile analysis is not only an appropriate approach to study related genome function but also is informative to analyze post-translational modification, protein-protein interactions, protein regulation mechanisms, and metabolic networks [71–73]. Therefore, proteomics is a powerful tool to identify genes responsible for stress tolerance. Moreover, proteome identification and physiology analysis could provide information to detect genome, stress, or term-related genes or potential biomarkers for a better description level of stress tolerance in each genotype [74].

Mass spectrometry assists with chromatographic instrumentation, and electrophoretic techniques are the primary method for proteome identification and quantification [75]. The application of advanced proteomics technologies like isobaric tags labeling allows characterizing a more significant number of proteins with lower abundance [76].

3.4.2 Proteome alteration under abiotic stress

Plant response to stress in four phases depends on the severity, duration, or recovery process, including the alarm, acclimation, resistance, and exhaustion phases. In the aspect of proteome level, the alarm phase involves modification in signaling-related proteins [77, 78]. Tolerant genotypes shift to the acclimation phase by activating energy and protein metabolism due to the high demand of energy production for enzymes activity. Glycolysis is one of the carbohydrate catabolism that upregulated under the acclimation process in contrast with enzymes related to the biosynthesis of energy-rich compounds like sucrose synthase [77, 79]. Protein metabolism defines a balance between protein biosynthesis and degradation to keep protein homeostasis in plants. The acclimation phase considers as one of the significant differences in proteome changes across stress-tolerant and stress-sensitive genotypes.

There are two main categories of proteins that participate in plant response to stress; the first group contains functional proteins. Late embryogenesis abundant (LEA) protein family plays a crucial role in tolerance mechanism in plants and found increasing in abundant under drought, heat, salinity, cold, and mechanical wounding [80–84]. Dehydrins are one group of LEA proteins whose overexpression under abiotic stress acclimate plants to abiotic stress, though their exact function is still unclear [85]. Another group of functional proteins is the heat shock protein (HSP) family and chaperones. These proteins are responsible for protecting proteins from aggregation and misfolding and consequently avoiding imbalances in cellular homeostasis. Many reports show that HSPs and chaperons are crucial in plant tolerance after exposure to abiotic stresses like extreme temperatures and drought stresses [86].

Enzymes cooperate in plant tolerance by two mechanisms, including energy production, as discussed before, and ascorbate peroxidase (APX) and catalase are two examples of proteins correlated to detoxification of reactive oxygen species (ROS) in response to abiotic stress like low-temperature conditions [87]. For example, accumulation of H_2O_2 as signal transduction molecule in response to abiotic stress, triggers to cellular damage, and inhibition of photosynthesis [88].

In response to different abiotic stress in plants, stress-induced signal transduction pathways initiate by various signaling receptors includes reactive oxygen species (ROS) [89–91]. Most dominating ROS include hydrogen peroxide (H_2O_2), hydroxyl radical (OH^*), singlet oxygen ($1O_2$), superoxide radical ($O_2{}^*{}_-$), etc. [92] that induced ROS overproduction serves various signaling cascades to regulate stress responses in plants like acclimation or defense by activating downstream metabolic pathways [90, 91, 93, 94]. Studies have found that ROS plays pivotal role mainly as primary signals under unfavorable conditions and interact with other signaling molecules like calcium, MAPKs, plant hormones, and transcription factors [95]. Oxidative stress is the consequence of any misbalance between ROS production and scavenging in plants, lead to proteins function and structure altering and damaging DNA, RNA, and other molecules [96–98].

In addition to ROS, phytohormones are activated under unfavorable condition by modulating stress-adaptive signaling cascades. Abscisic acid (ABA) regulates various physiological processes ranging from stomatal opening to protein storage and coordinate complex stress-adaptive signaling cascades. ABA is also an important messenger that acts as a critical regulator in activating plant cellular adaptation to different environmental stress conditions [99, 100].

The other functional proteins related to stress tolerance involve ion transporter and membrane proteins that help keep membrane stability by an incredible increase. V-ATPase and channel proteins (NHX-1) increased under salinity in plants [85, 101].

The second group of proteins activated under stress is regulatory proteins, namely, photosynthesis-related proteins, Transcription factors (TF), kinases, phosphatases, and signaling proteins [85]. The photosynthesis mechanism has a very crucial role in plant reaction to stress. Plant decreases photosynthesis-related proteins abundant as try to slow down the growth pace to avoid death happening and closing stomata to avoid water loss, such as RubisCO and chlorophyll a-b binding proteins in both photosystems (PS) I and II found downregulated under low temperature and flood stress [102]. Transcription factors (TF) generally regulate genes expression under stress with the ability of binding specific sequences to these genes and lead tolerance in plants exposed to stress [103]. Mitogen-activated protein kinase (MAPK) is an example of kinases proteins with a particular function in regulating plant responses to unfavorable environmental conditions [104, 105]. On the other hand, kinases coordinate in stress tolerance by the phosphorylation mechanism. Several kinases are found with this function, such as CDPK, MAPK, and SnRK2 [106]. The major subcellular organelles, whose functions get affected under abiotic stress, are the nucleus, mitochondria, chloroplasts, peroxisomes, plasma membrane, and cell wall. Most of these organelles have the potential to become a source of ROS.

3.4.2.1 Proteome response to salt stress

Salinity stress represents enhanced levels of salt ions in soil water solution. As a consequence of enhanced ion levels, decreased soil water potential reveals a so-called osmotic effect on plant cells [78].

Increasing salt (mainly Na^+) accumulation which is known as salt stress in plants, leads to several metabolite modifications; however, the plant species, the age of the plant, and Na^+ concentration are three main factors affecting the stress severity. For example, by increasing Na^+ concentration, salt stress ranges from osmotic stress to osmotic shock [107]. In the short term, high Na^+ and Cl^- concentration causes osmotic stress by misbalance in intracellular ion homeostasis due to decreasing K^+ transport with excess toxic intracellular Na^+ ions in the cytosol; therefore, plants exposes to ionic stress [108, 109].

Osmotic stress affects plant cellular metabolism in different ways like membrane fluidity, production, and accumulation of reactive oxygen species (ROS) and further oxidative stress happening, photosynthesis malfunction by stomatal conductance decreasing [108–110]. The K^+/Na^+ discrimination ratio causes inhibiting K^+ ion uptaking as a vital factor related to many growing and development functions in plants indirectly leads to plant death [111]. To balance ionic homeostasis, an increase in the concentration of cytosolic Ca^{2+} activates some proteins like antiporters and enzymes involved in the transport of ions and phospholipases [112]. Induced accumulation of Ca^{2+} activates H^+-ATPases as an ion-transporting protein to maintain cytosolic ion homeostasis. On the other hand, the elevated activity of H^+-ATPases is vital to avoid electrochemical gradient misbalance and maintain cellular pH homeostasis [113, 114].

Osmotic stress is another consequence of a high concentration of Na^+ in soil, which could be considered a primary signal of ionic stress. Plants sense the osmotic stress by losing leaf water due to high salinity and water capacity absorption decreasing. Therefore, salinity is known as hyperosmotic stress [108, 115]. Na^+ uptaking to plant roots happens via apoplastic or symplastic routes through Na^+ transport transmembrane proteins and Na^+/H^+ exchangers.

Oxidative homeostasis is misbalanced due to photosynthetic activity and energy metabolism reduction, which induce the accumulation of ROS. Intracellular ROS is a critical signaling molecule promoting oxidative stress induction [41, 116]. Excess ROS accumulation, recognized as a marker of oxidative stress, may be removed by ROS scavengers. Two main groups have been found responsible for oxidative stress response in plants like enzymatic or non-enzymatic antioxidants. Antioxidant enzymes are identified by ameliorating some antioxidative enzyme activities, while other antioxidant mechanisms could play more effective participation in oxidative response [117, 118]. Recent studies on crop plants have identified the following five main groups of proteins that present differential abundance and are directly related to salt stress response mechanisms: (I) heat shock proteins (HSPs), (II) late embryogenesis abundant proteins (LEA proteins), (III) osmolyte and flavonoids biosynthetic enzymes, (IV) proteins involved in carbon, photosynthesis and energy metabolism like rubisco activase, kinases, and oxygen-evolving enhancer protein 2 (OEE 2), and (V) enzyme scavengers of ROS such as catalase, peroxidase, and GST [86, 102, 113, 117, 119].

3.4.2.2 Proteome response to drought stress

Drought means a decreased soil water potential which causes reduced water uptake by roots. Plant response at the molecular level lies in osmotic adjustment, that is, a decrease in osmotic potential of cell cytoplasm due to an enhanced accumulation of several osmolytes and hydrophilic proteins. The primary signal caused by drought is hyperosmotic stress, which is often referred to simply as osmotic stress because a hypo-osmotic condition typically is not a significant problem for plant cells. In leaves, drought leads to stomatal closure associated with reduced CO_2 uptake resulting in imbalances between photosynthetic electron transport processes and carbon assimilation. As a consequence, cellular dehydration is also associated with enhanced ROS formation [120].

Water losing causes decreasing in leaf water potential in contrast with cytoplasmic components concentration and extracellular matrices viscosity increase [121, 122]. In response to drought stress, CO_2 production declines due to stomatal closing. On the other hand, photosynthesis may be known as the first and most significant function negatively affected by water deficiency stress [123, 124].

Decreasing in internal CO_2 concentration impacts on Calvin cycle by disturbing carbon fixation accompanied with a proportional reduction in the activity of various related enzymes like fructose-1,6-biphosphate phosphatase (FBPase), ribulose-1,5-bisphosphate carboxylase-oxygenase (Rubisco), and phosphoribulokinase (PRK). Consequently, glycerate-3-phosphate depletion and $NADP^+$ production decline lead to carbohydrate formation decrease in the final step of the Calvin cycle. Moreover, $NADP^+$ is the primary electron acceptor in photosystem I (PSI), and therefore O_2 production, lowered as the final production of photosynthesis in plants and excessive generation of reactive oxygen species (ROS) occurs [125–128]. Moreover, reducing ATP production in response to drought stress is associated with photosynthesis function by negatively affecting the Calvin cycle.

A coordinated down-regulation of the Calvin cycle genes with a decline in carbon fixation in plants exposed to drought stress reduces the energy-wasting for unnecessary biomolecule synthesis through the lower level of carbohydrate production [129].

Under drought stress, due to low CO_2:O_2 ratio ribulose-1,5-bisphosphate carboxylase-oxygenase (RuBisCO) enzyme oxygenates ribulose 1,5-bisphosphate (RuBP) and causes photorespiration (C_2 photosynthesis) cycle to function as a carbon recovery system which is correlated with the Calvin cycle [130].

The Krebs cycle (TCA cycle) is one of the primary energy sources for cells and essential for aerobic respiration. There are contradictory reports for effects of drought stress on the Krebs cycle and its intermediate such as PDHA1, NADP-ME, and α-ketoglutaric acid [131–133].

Another cycle affected by drought stress is the ascorbate-glutathione (ASA-GSH) cycle detoxifies waste hydrogen peroxide (H_2O_2) to H_2O by using antioxidant metabolites like ascorbate, glutathione, and NADPH and the enzymes linking these metabolites [134, 135].

3.4.2.3 Proteome response to temperature stress

Temperature stress (both heat and cold) leads to an imbalance between photosynthetic electron transport processes and carbon assimilation processes resulting in enhanced photoinhibition and thermal energy dissipation [136] and is linked to plant metabolism and performance [137]. The synthesis and accumulation of heat-shock proteins (HSPs) is a prompt response after exposure to extreme temperature stress treatment. It is considered one of the most critical adaptive strategies to overcome the deleterious effects of temperature fluctuation stress [138, 139]. Most HSPs are molecular chaperones involved in protein stabilization and signal transduction during heat stress [140, 141].

Low temperatures (cold and frost) can induce inhibition of water uptake and indirectly result in osmotic stress in cells [142]. Consequently, decreased kinetics of biomolecules leading to reduced cell membrane fluidity and a decreased rate of enzymatic reactions. Frost stress particularly leads to ice crystals in soil, resulting in cellular dehydration [78].

High temperature (heat) stress as enhanced temperature causes enhanced kinetics of biomolecules leading to an enhanced risk of protein misfolding.

Thus, plant response includes induction of several heat-shock transcription factors (HSFs) and downstream heat-shock proteins (HSPs). Moreover, heat causes enhanced water evaporation from the soil surface and enhanced leaf transpiration, thus usually resulting in water deficit under field conditions. The heat thus also causes dehydration stress and oxidative stress [136].

4. Conclusions

Plant production in two aspects of quality and quantity plays a key role in food safety and security impressed with global environmental stresses increasing with emphasizing on study to detect plant alterations in response to abiotic stresses. Different levels of plants sense stresses from morphological to molecular aspects however finding deep mechanisms and functions affected by stresses would help to find related genes and biomarkers more accurately. We reviewed the significance of agriculture for our future in addition to its mutual relationship with environment. Climate change as the most effective parameter correlated with food production, is susceptible to impact negatively on both food safety and food security.

Analyzing the proteome profile of plants exposing various types of abiotic stresses including salt, drought and extreme temperature is helpful to find shared and unique mechanisms related to abiotic stresses in addition to finding solution useful for crop production increasing and sustainability.

Author details

Fatemeh Habibpourmehraban
Macquarie University, Sydney, Australia

*Address all correspondence to: fatemeh.habibpour-mehraban@students.mq.edu.au

References

[1] Helland J, Sørbø GM. Food securities and social conflict. CMI Report. 2014;**2014**(1):25. DOI: 10.1073/pnas.10055739107a

[2] Vågsholm I, Arzoomand NS, Boqvist S. Food security, safety, and sustainability-getting the trade-offs right. Frontiers in Sustainable Food Systems. 2020;**4**:16. DOI: 10.3389/fsufs.2020.00016

[3] Porter JR, Xie L, Challinor AJ, Cochrane K, Howden SM, Iqbal MM, et al. Food security and food production systems. In: Field CB, Barros VR, Dokken DJ, Mach KJ, Mastrandrea MD, Bilir TE, et al. editors. Climate Change 2014: Impacts, Adaptation, and Vulnerability. Part A: Global and Sectoral Aspects. Contribution of Working Group II to the Fifth Assessment Report of the Intergovernmental Panel on Climate Change. Cambridge, United Kingdom and New York, NY, USA: Cambridge University Press; 2014. pp. 485-533

[4] UNFCCC. Conference of the Parties (COP). Adoption of the Paris agreement. Revision/CP. 21: CP/L 9/Rev 1. 2015

[5] Dos Reis SP, Lima AM, De Souza CRB. Recent molecular advances on downstream plant responses to abiotic stress. International Journal of Molecular Sciences. 2012;**13**:8628-8647. DOI: 10.3390/ijms13078628

[6] Mantri N, Patade V, Penna S, Ford R, Pang E. Abiotic stress responses in plants: Present and future. In: Ahmad P, Prasad M, editors. Abiotic Stress Responses in Plants. New York, United States: Springer; 2012. pp. 1-19. DOI: 10.1007/978-1-4614-0634-1_1

[7] Vermeulen SJ, Campbell BM, Ingram JS. Climate change and food systems. Annual Review of Environment and Resources. 2012;**37**:195-222. DOI: 10.1146/annurev-environ-020411-130608

[8] Mekonnen M, Hoekstra AY. The green, blue and grey water footprint of crops and derived crops products. Hydrology and Earth System Sciences. 2011;**15**:1577-1600. DOI: 10.5194/hess-15-1577-2011

[9] Lobell DB, Gourdji SM. The influence of climate change on global crop productivity. Plant Physiology. 2012;**160**:1686-1697. DOI: 10.1104/pp.112.208298

[10] Field CB, Barros VR. Climate Change 2014: Impacts, Adaptation and Vulnerability: Regional Aspects. 6th ed. Cambridge, United Kingdom and New York, NY, USA: Cambridge University Press; 2014. pp. 1-32. DOI: 10.1017/CBO9781107415386

[11] Tuomisto H, Scheelbeek P, Chalabi Z, Green R, Smith R, Haines A, et al. Effects of environmental change on agriculture, nutrition and health: A framework with a focus on fruits and vegetables. Wellcome Open Research. 2017;**2**:21. DOI: 10.12688/wellcomeopenres.11190.2

[12] Freedman B. Environmental Stressors. Environmental Science. Halifax, Nova Scotia, Canada: Dalhousie University Libraries. A Creative Commons Attribution-NonCommercial 4.0 International License; 2018. pp. 431-467

[13] Pereira A. Plant abiotic stress challenges from the changing environment. Frontiers in Plant Science. 2016;**7**:1123. DOI: 10.3389/fpls.2016.01123

[14] Gull A, Lone AA, Wani NUI. Biotic and abiotic stresses in plants. In: Abiotic

and Biotic Stress in Plants. Alexandre Bosco de Oliveira: IntechOpen; 2019. pp. 1-19. DOI: 10.5772/intechopen.85832

[15] Des Marais DL, Hernandez KM, Juenger TE. Genotype-by-environment interaction and plasticity: Exploring genomic responses of plants to the abiotic environment. Annual Review of Ecology, Evolution, and Systematics. 2013;**44**:5-29. DOI: 10.1146/annurev-ecolsys-110512-135806

[16] Verma S, Nizam S, Verma PK. Biotic and abiotic stress signaling in plants. Stress Signaling in Plants: Genomics and Proteomics Perspective. 2013;**1**:25-49. DOI: 10.1007/978-1-4614-6372-6_2

[17] Haggag WM, Abouziena H, Abd-El-Kreem F, El Habbasha S. Agriculture biotechnology for management of multiple biotic and abiotic environmental stress in crops. Journal of Chemical Pharmaceutical Research. 2015;7:882-889

[18] De Storme N, Geelen D. The impact of environmental stress on male reproductive development in plants: Biological processes and molecular mechanisms. Plant, Cell & Environment. 2014;**37**:1-18. DOI: 10.1111/pce.12142

[19] Hammond-Kosack K, Jones JDG. Responses to plant pathogens. In: Jones RL, Buchanan BB, Gruissem W, editors. Biochemistry and Molecular Biology of Plants. 2nd ed. Chichester, West Sussex: John Wiley & Sons; 2015. pp. 984-1050

[20] Sah SK, Reddy KR, Li J. Abscisic acid and abiotic stress tolerance in crop plants. Frontiers in Plant Science. 2016;7:571. DOI: 10.3389/fpls.2016.00571

[21] Peck S, Mittler R. Plant signaling in biotic and abiotic stress. Journal of Experimental Botany. 2020;**71**(5): 1649-1651. DOI: 10.1093/jxb/eraa051

[22] Lamers J, Van Der Meer T, Testerink C. How plants sense and respond to stressful environments. Plant Physiology. 2020;**182**:1624-1635. DOI: 10.1104/pp.19.01464

[23] Nejat N, Mantri N. Plant immune system: Crosstalk between responses to biotic and abiotic stresses the missing link in understanding plant defense. Current Issues in Molecular Biology. 2017;**23**:1-16. DOI: 10.21775/cimb.023.001

[24] Saijo Y, Epi L. Plant immunity in signal integration between biotic and abiotic stress responses. The New Phytologist. 2020;**225**:87-104. DOI: 10.1111/nph.15989

[25] Bates B, Kundzewicz Z, Wu S, Palutikof J. Climate change and water. In: Technical Paper of the Intergovernmental Panel on Climate Change. Geneva: IPCC Secretariat; 2008. p. 210

[26] Lee H. Intergovernmental Panel on Climate Change. 2007

[27] Mittler R, Blumwald E. Genetic engineering for modern agriculture: Challenges and perspectives. Annual Review of Plant Biology. 2010;**61**:443-462. DOI: 10.1146/annurev-arplant-042809-112116

[28] Sehgal A, Sita K, Siddique KHM, Kumar R, Bhogireddy S, Varshney RK, et al. Drought or/and heat-stress effects on seed filling in food crops: Impacts on functional biochemistry, seed yields, and nutritional quality. Frontiers in Plant Science. 2018;**9**:1705. DOI: 10.3389/fpls.2018.01705

[29] Sulmon C, Van Baaren J, Cabello-Hurtado F, Gouesbet G, Hennion F, Mony C, et al. Abiotic stressors and stress responses: What commonalities appear between species across biological organization levels? Environmental Pollution. 2015;**202**:66-77. DOI: 10.1016/j.envpol.2015.03.013

[30] Vaughan MM, Block A, Christensen SA, Allen LH, Schmelz EA.

The effects of climate change associated abiotic stresses on maize phytochemical defenses. Phytochemistry Reviews. 2018;**17**:37-49. DOI: 10.1007/s11101-017-9508-2

[31] Zafar SA, Noor MA, Waqas MA, Wang X, Shaheen T, Raza M, et al. Temperature extremes in cotton production and mitigation strategies. In: Mehboob-Ur-Rahman, Zafar Y, editors. Past, Present and Future Trends in Cotton Breeding. [Internet]. London: IntechOpen; 2018. 184 p. Available from: https://www.intechopen.com/books/6328 [cited 2022 Mar 10]

[32] Lobell DB, Burke MB, Tebaldi C, Mastrandrea MD, Falcon WP, Naylor RL. Prioritizing climate change adaptation needs for food security in 2030. Science. 2008;**319**(5863):607-610. DOI: 10.1126/science.1152339

[33] Wang W, Vinocur B, Altman A. Plant responses to drought, salinity and extreme temperatures: Towards genetic engineering for stress tolerance. Planta. 2003;**218**:1-14. DOI: 10.1007/s00425-003-1105-5

[34] FAO. The State of the World's Land and Water Resources for Food and Agriculture (SOLAW)—Managing Systems at Risk. Rome and Earthscan, London: Food and Agriculture Organization of the United Nations; 2011

[35] Adams RM, Hurd BH, Lenhart S, Leary N. Effects of global climate change on agriculture: An interpretative review. Climate Research. 1998;**11**:19-30. DOI: 10.3354/cr011019

[36] National Research Council. America's Climate Choices. Washington, DC: The National Academies Press; 2011. DOI: 10.17226/12781

[37] Dixit S, Singh A, Kumar A. Rice breeding for high grain yield under drought: A strategic solution to a complex problem. International Journal of Agronomy. 2014;**2014**. DOI: 10.1155/2014/863683

[38] Al-Karaki G. Growth, sodium, and potassium uptake and translocation in salt stressed tomato. Journal of Plant Nutrition. 2000;**23**:369-379. DOI: 10.1080/01904160009382023

[39] Pons R, Cornejo M-J, Sanz A. Differential salinity-induced variations in the activity of H^+-pumps and Na^+/H^+ antiporters that are involved in cytoplasm ion homeostasis as a function of genotype and tolerance level in rice cell lines. Plant Physiology and Biochemistry. 2011;**49**:1399-1409. DOI: 10.1016/j.plaphy.2011.09.011

[40] Shahbaz M, Ashraf M. Improving salinity tolerance in cereals. Critical Reviews in Plant Sciences. 2013;**32**:237-249. DOI: 10.1080/07352689.2013.758544

[41] Munns R, Tester M. Mechanisms of salinity tolerance. Annual Review of Plant Biology. 2008;**59**:651-681. DOI: 10.1146/annurev.arplant.59.032607.092911

[42] Ashraf M, Foolad M. Crop breeding for salt tolerance in the era of molecular markers and marker-assisted selection. Plant Breeding. 2013;**132**:10-20. DOI: 10.1111/pbr.12000

[43] Inal A, Gunes A. Interspecific root interactions and rhizosphere effects on salt ions and nutrient uptake between mixed grown peanut/maize and peanut/barley in original saline–sodic–boron toxic soil. Journal of Plant Physiology. 2008;**165**:490-503. DOI: 10.1016/j.jplph.2007.01.016

[44] Lang T, Deng S, Zhao N, Deng C, Zhang Y, Zhang Y, et al. Salt-sensitive signaling networks in the mediation of K^+/Na^+ homeostasis gene expression in *Glycyrrhiza uralensis* roots. Frontiers in Plant Science. 2017;**8**:1403. DOI: 10.3389/fpls.2017.01403

[45] Hasanuzzaman M, Raihan MRH, Masud AAC, Rahman K, Nowroz F, Rahman M, et al. Regulation of reactive oxygen species and antioxidant defense in plants under salinity. International Journal of Molecular Sciences. 2021;**22**(17):9326. DOI: 10.3390/ijms22179326

[46] Suzuki K, Nagasuga K, Okada M. The chilling injury induced by high root temperature in the leaves of rice seedlings. Plant & Cell Physiology. 2008;**49**:433-442. DOI: 10.1093/pcp/pcn020

[47] Zhu J, Dong C-H, Zhu J-K. Interplay between cold-responsive gene regulation, metabolism and RNA processing during plant cold acclimation. Current Opinion in Plant Biology. 2007;**10**:290-295. DOI: 10.1016/j.pbi.2007.04.010

[48] Asthir B. Mechanisms of heat tolerance in crop plants. Journal of Plant Interactions. 2015;**10**:1-21. DOI: 10.1080/17429145.2015.1067726

[49] Devasirvatham V, Tan DK. Impact of high temperature and drought stresses on chickpea production. Agronomy. 2018;**8**:145. DOI: 10.3390/agronomy8080145

[50] Hasanuzzaman M, Nahar K, Alam MM, Roychowdhury R, Fujita M. Physiological, biochemical, and molecular mechanisms of heat stress tolerance in plants. International Journal of Molecular Sciences. 2013;**14**(5):9643-9684. DOI: 10.3390/ijms14059643

[51] Takahashi D, Li B, Nakayama T, Kawamura Y, Uemura M. Plant plasma membrane proteomics for improving cold tolerance. Frontiers in Plant Science. 2013;**4**:90. DOI: 10.3389/fpls.2013.00090

[52] Proshad R, Kormoker T, Mursheed N, Islam MM, Bhuyan MI, Islam MS, et al. Heavy metal toxicity in agricultural soil due to rapid industrialization in Bangladesh: A review. International Journal of Advanced Geosciences. 2018;**6**:83-88. DOI: 10.14419/ijag.v6i1.9174

[53] Minhas PS, Rane J, Pasala RK. Abiotic stress management for resilient agriculture. In: Minhas PS, Rane J, Pasala RK, editors. 1 st ed. Singapore: Springer; 2017. DOI: 10.1007/978-981-10-5744-1

[54] Khan ZI, Ugulu I, Sahira S, Ahmad K, Ashfaq A, Mehmood N, et al. Determination of toxic metals in fruits of *Abelmoschus esculentus* grown in contaminated soils with different irrigation sources by spectroscopic method. International Journal of Environmental Research. 2018;**12**:503-511. DOI: 10.30848/PJB2020-1(12)

[55] Narendrula-Kotha R, Theriault G, Mehes-Smith M, Kalubi K, Nkongolo K. Metal toxicity and resistance in plants and microorganisms in terrestrial ecosystems. Reviews of Environmental Contamination and Toxicology. 2019;**249**:1-27. DOI: 10.1007/398_2018_22

[56] Schutzendubel A, Polle A. Plant responses to abiotic stresses: Heavy metal-induced oxidative stress and protection by mycorrhization. Journal of Experimental Botany. 2002;**53**: 1351-1365

[57] Hall J. Cellular mechanisms for heavy metal detoxification and tolerance. Journal of Experimental Botany. 2002;**53**(366):1-11

[58] Fatima U, Senthil-Kumar M. Tissue water status and bacterial pathogen infection: How they are correlated? In: Senthil-Kumar M, editor. Plant Tolerance to Individual and Concurrent Stresses. New Delhi: Springer; 2017. pp. 165-178. DOI: 10.1007/978-81-322-3706-8_11

[59] Maiti R, Satya P. Research advances in major cereal crops for adaptation to

abiotic stresses. GM Crops & Food. 2014;**5**:259-279. DOI: 10.4161/21645698.2014.947861

[60] Nguyen D, Rieu I, Mariani C, van Dam NM. How plants handle multiple stresses: Hormonal interactions underlying responses to abiotic stress and insect herbivory. Plant Molecular Biology. 2016;**91**:727-740. DOI: 10.1007/s11103-016-0481-8

[61] Grimmer MK, John Foulkes M, Paveley ND. Foliar pathogenesis and plant water relations: A review. Journal of Experimental Botany. 2012;**63**:4321-4331. DOI: 10.1093/jxb/ers143

[62] Zhang H, Sonnewald U. Differences and commonalities of plant responses to single and combined stresses. The Plant Journal. 2017;**90**:839-855. DOI: 10.1111/tpj.13557

[63] RoyChoudhury A, Roy C, Sengupta DN. Transgenic tobacco plants overexpressing the heterologous lea gene Rab16A from rice during high salt and water deficit display enhanced tolerance to salinity stress. Plant Cell Reports. 2007;**26**:1839-1859. DOI: 10.1007/s00299-007-0371-2

[64] Li J, Cang Z, Jiao F, Bai X, Zhang D, Zhai R. Influence of drought stress on photosynthetic characteristics and protective enzymes of potato at seedling stage. Journal of the Saudi Society of Agricultural Sciences. 2017;**16**:82-88. DOI: 10.1016/j.jssas.2015.03.001

[65] Zhang Z-F, Li Y-Y, Xiao B-Z. Comparative transcriptome analysis highlights the crucial roles of photosynthetic system in drought stress adaptation in upland rice. Scientific Reports. 2016;**6**:1-13. DOI: 10.1038/srep19349

[66] Hahn A, Kilian J, Mohrholz A, Ladwig F, Peschke F, Dautel R, et al. Plant core environmental stress response genes are systemically coordinated during abiotic stresses. International Journal of Molecular Sciences. 2013;**14**:7617-7641. DOI: 10.3390/ijms14047617

[67] Mohan-Raju B, Paramanantham A, Ramegowda V, Udayakumar M, Senthil-Kumar M, Ramu VS. Transcriptome analysis of sunflower genotypes with contrasting oxidative stress tolerance reveals individual-and combined-biotic and abiotic stress tolerance mechanisms. PLoS One. 2016;**11**(6):e0157522. DOI: 10.1371/journal.pone.0157522

[68] Pandey P, Ramegowda V, Senthil-Kumar M. Shared and unique responses of plants to multiple individual stresses and stress combinations: Physiological and molecular mechanisms. Frontiers in Plant Science. 2015;**6**:723. DOI: 10.3389/fpls.2015.00723

[69] Roychoudhury A, Basu S, Sarkar SN, Sengupta DN. Comparative physiological and molecular responses of a common aromatic indica rice cultivar to high salinity with non-aromatic indica rice cultivars. Plant Cell Reports. 2008;**27**:1395-1410. DOI: 10.1007/s00299-008-0556-3

[70] Cohen SP, Leach JE. Abiotic and biotic stresses induce a core transcriptome response in rice. Scientific Reports. 2019;**9**:6273. DOI: 10.1038/s41598-019-42731-8

[71] Komatsu S, Hossain Z. Organ-specific proteome analysis for identification of abiotic stress response mechanism in crop. Frontiers in Plant Science. 2013;**4**:71. DOI: 10.3389/fpls.2013.00071

[72] Komatsu S, Konishi H, Shen S, Yang G. Rice proteomics: A step toward functional analysis of the rice genome. Molecular & Cellular Proteomics. 2003;**2**:2-10. DOI: 10.1074/mcp.r200008-mcp200

[73] Komatsu S, Yano H. Update and challenges on proteomics in rice. Proteomics. 2006;**6**:4057-4068. DOI: 10.1002/pmic.200600012

[74] Komatsu S. Research on the rice proteome: The contribution of proteomics technology in the creation of abiotic stress-tolerant plants. Rice. 2008;**1**:154-165. DOI: 10.1007/s12284-008-9013-8

[75] Glinski M, Weckwerth W. The role of mass spectrometry in plant systems biology. Mass Spectrometry Reviews. 2006;**25**:173-214. DOI: 10.1002/mas.20063

[76] Agrawal GK, Rakwal R. Rice proteomics: A move toward expanded proteome coverage to comparative and functional proteomics uncovers the mysteries of rice and plant biology. Proteomics. 2011;**11**:1630-1649. DOI: 10.1002/pmic.201000696

[77] Kosová K, Vítámvás P, Prášil IT. Proteomics of stress responses in wheat and barley—Search for potential protein markers of stress tolerance. Frontiers in Plant Science. 2014;**5**:711. DOI: 10.3389/fpls.2014.00711

[78] Kosová K, Vítámvás P, Urban MO, Prášil IT, Renaut J. Plant abiotic stress proteomics: The major factors determining alterations in cellular proteome. Frontiers in Plant Science. 2018;**9**:122. DOI: 10.3389/fpls.2018.00122

[79] Vítámvás P, Prášil IT, Kosová K, Planchon S, Renaut J. Analysis of proteome and frost tolerance in chromosome 5A and 5B reciprocal substitution lines between two winter wheats during long-term cold acclimation. Proteomics. 2012;**12**:68-85. DOI: 10.1002/pmic.201000779

[80] Kosová K, Holková L, Prášil IT, Prášilová P, Bradáčová M, Vítámvás P, et al. Expression of dehydrin 5 during the development of frost tolerance in barley (*Hordeum vulgare*). Journal of Plant Physiology. 2008;**165**:1142-1151. DOI: 10.1016/j.jplph.2007.10.009

[81] Sarhadi E, Mahfoozi S, Hosseini SA, Salekdeh GH. Cold acclimation proteome analysis reveals close link between the up-regulation of low-temperature associated proteins and vernalization fulfillment. Journal of Proteome Research. 2010;**9**:5658-5667. DOI: 10.1021/pr100475r

[82] Vágújfalvi A, Galiba G, Cattivelli L, Dubcovsky J. The cold-regulated transcriptional activator Cbf3 is linked to the frost-tolerance locus Fr-A2 on wheat chromosome 5A. Molecular Genetics and Genomics. 2003;**269**:60-67. DOI: 10.1007/s00438-003-0806-6

[83] Vágújfalvi A, Galiba G, Dubcovsky J, Cattivelli L. Two loci on wheat chromosome 5A regulate the differential cold-dependent expression of the cor14b gene in frost-tolerant and frost-sensitive genotypes. Molecular and General Genetics MGG. 2000;**263**:194-200. DOI: 10.1007/s004380051160

[84] Vítámvás P, Saalbach G, Prášil IT, Čapková V, Opatrná J, Ahmed J. WCS120 protein family and proteins soluble upon boiling in cold-acclimated winter wheat. Journal of Plant Physiology. 2007;**164**:1197-1207. DOI: 10.1016/j.jplph.2006.06.011

[85] Roychoudhury A, Chakraborty M. Biochemical and molecular basis of varietal difference in plant salt tolerance. Annual Research & Review in Biology. 2013:422-454

[86] Rodziewicz P, Swarcewicz B, Chmielewska K, Wojakowska A, Stobiecki M. Influence of abiotic stresses on plant proteome and metabolome changes. Acta Physiologiae Plantarum. 2014;**36**:1-19. DOI: 10.1007/s11738-013-1402-y

[87] Hashimoto M, Komatsu S. Proteomic analysis of rice seedlings during cold stress. Proteomics. 2007;7:1293-1302. DOI: 10.1002/pmic.200600921

[88] Baxter A, Mittler R, Suzuki N. ROS as key players in plant stress signaling. Journal of Experimental Botany. 2014;**65**(5):1229-1240. DOI: 10.1093/jxb/ert375

[89] Devireddy AR, Zandalinas SI, Fichman Y, Mittler R. Integration of reactive oxygen species and hormone signaling during abiotic stress. The Plant Journal. 2021;**105**:459-476. DOI: 10.1111/tpj.15010

[90] Mittler R. ROS are good. Trends in Plant Science. 2017;**22**:11-19. DOI: 10.1016/j.tplants.2016.08.002

[91] Waszczak C, Carmody M, Kangasjärvi J. Reactive oxygen species in plant signaling. Annual Review of Plant Biology. 2018;**69**:209-236. DOI: 10.1146/annurev-arplant-042817-040322

[92] Mittler R. Abiotic stress, the field environment and stress combination. Trends in Plant Science. 2006;**11**:15-19. DOI: 10.1016/j.tplants.2005.11.002

[93] Chan Z, Yokawa K, Kim W-Y, Song C-P. Editorial: ROS regulation during plant abiotic stress responses. Frontiers in Plant Science. 2016;7:1536. DOI: 10.3389/fpls.2016.01536

[94] Noctor G, Reichheld J-P, Foyer CH. ROS-related redox regulation and signaling in plants. Cell & Developmental Biology Seminar. 2018;**80**:3-12. DOI: 10.1016/j.semcdb.2017.07.013

[95] Sewelam N, Kazan K, Schenk PM. Global plant stress signaling: Reactive oxygen species at the cross-road. Frontiers in Plant Science. 2016;7:187. DOI: 10.3389/fpls.2016.00187

[96] Dietz K-J. Thiol-based peroxidases and ascorbate peroxidases: Why plants rely on multiple peroxidase systems in the photosynthesizing chloroplast? Molecules and Cells. 2016;**39**(1):20-25. DOI: 10.14348/molcells.2016.2324

[97] Druege U, Franken P, Hajirezaei MR. Plant hormone homeostasis, signaling, and function during adventitious root formation in cuttings. Frontiers in Plant Science. 2016;7:381. DOI: 10.3389/fpls.2016.00381

[98] Nath M, Bhatt D, Prasad R, Gill SS, Anjum NA, Tuteja N. Reactive oxygen species generation-scavenging and signaling during plant-arbuscular mycorrhizal and *Piriformospora indica* interaction under stress condition. Frontiers in Plant Science. 2016;7:1574. DOI: 10.3389/fpls.2016.01574

[99] Danquah A, de Zelicourt A, Colcombet J, Hirt H. The role of ABA and MAPK signaling pathways in plant abiotic stress responses. Biotechnology Advances. 2014;**32**:40-52. DOI: 10.1016/j.biotechadv.2013.09.006

[100] Muñoz-Espinoza VA, López-Climent MF, Casaretto JA, Gómez-Cadenas A. Water stress responses of tomato mutants impaired in hormone biosynthesis reveal abscisic acid, jasmonic acid and salicylic acid interactions. Frontiers in Plant Science. 2015;**6**:997. DOI: 10.3389/fpls.2015.00997

[101] Peng Z, Wang M, Li F, Lv H, Li C, Xia G. A proteomic study of the response to salinity and drought stress in an introgression strain of bread wheat. Molecular & Cellular Proteomics. 2009;**8**:2676-2686. DOI: 10.1074/mcp.M900052-MCP200

[102] Li X, Cai J, Liu F, Dai T, Cao W, Jiang D. Physiological, proteomic and transcriptional responses of wheat to combination of drought or waterlogging with late spring low temperature. Functional Plant Biology. 2014;**41**:690-703. DOI: 10.1071/FP13306

[103] Nakashima K, Ito Y, Yamaguchi-Shinozaki K. Transcriptional regulatory networks in response to abiotic stresses in *Arabidopsis* and grasses. Plant Physiology. 2009;**149**:88-95. DOI: 10.1104/pp.108.129791

[104] Meng X, Zhang S. MAPK cascades in plant disease resistance signaling. Annual Review of Phytopathology. 2013;**51**:245-266. DOI: 10.1146/annurev-phyto-082712-102314

[105] Tena G, Boudsocq M, Sheen J. Protein kinase signaling networks in plant innate immunity. Current Opinion in Plant Biology. 2011;**14**:519-529. DOI: 10.1016/j.pbi.2011.05.006

[106] Zhang X, Lu G, Long W, Zou X, Li F, Nishio T. Recent progress in drought and salt tolerance studies in Brassica crops. Breeding Science. 2014;**64**:60-73. DOI: 10.1270/jsbbs.64.60

[107] Shavrukov Y. Salt stress or salt shock: Which genes are we studying? Journal of Experimental Botany. 2013;**64**(1):119-127. DOI: 10.1093/jxb/ers316

[108] Gupta B, Huang B. Mechanism of salinity tolerance in plants: Physiological, biochemical, and molecular characterization. International Journal of Genomics. 2014;**2014**:18. Article ID 701596. DOI: 10.1155/2014/701596

[109] Munns R. Comparative physiology of salt and water stress. Plant, Cell & Environment. 2002;**25**:239-250. DOI: 10.1046/j.0016-8025.2001.00808.x

[110] Rahnama A, James RA, Poustini K, Munns R. Stomatal conductance as a screen for osmotic stress tolerance in durum wheat growing in saline soil. Functional Plant Biology. 2010;**37**: 255-263. DOI: 10.1071/FP09148

[111] James RA, Blake C, Byrt CS, Munns R. Major genes for Na^+ exclusion, Nax1 and Nax2 (wheat HKT1; 4 and HKT1; 5), decrease Na^+ accumulation in bread wheat leaves under saline and waterlogged conditions. Journal of Experimental Botany. 2011;**62**:2939-2947. DOI: 10.1093/jxb/err003

[112] Wang X. Plant phospholipases. Annual Review of Plant Biology. 2001;**52**:211-231. DOI: 10.1146/annurev.arplant.52.1.211

[113] de Abreu CEB, Araújo GS, Monteiro-Moreira ACO, Costa JH, Leite HB, Moreno FBMB, et al. Proteomic analysis of salt stress and recovery in leaves of *Vigna unguiculata* cultivars differing in salt tolerance. Plant Cell Reports. 2014;**33**:1289-1306. DOI: 10.1007/s00299-014-1616-5

[114] Halperin SJ, Lynch JP. Effects of salinity on cytosolic Na^+ and K^+ in root hairs of *Arabidopsis thaliana*: In vivo measurements using the fluorescent dyes SBFI and PBFI. Journal of Experimental Botany. 2003;**54**:2035-2043. DOI: 10.1093/jxb/erg219

[115] Munns R. Genes and salt tolerance: Bringing them together. The New Phytologist. 2005;**167**:645-663. DOI: 10.1111/j.1469-8137.2005.01487.x

[116] Gill SS, Tuteja N. Reactive oxygen species and antioxidant machinery in abiotic stress tolerance in crop plants. Plant Physiology and Biochemistry. 2010;**48**:909-930. DOI: 10.1016/j.plaphy.2010.08.016

[117] Fini A, Brunetti C, Di Ferdinando M, Ferrini F, Tattini M. Stress-induced flavonoid biosynthesis and the antioxidant machinery of plants. Plant Signaling & Behavior. 2011;**6**:709-711. DOI: 10.4161/psb.6.5.15069

[118] Mostek A, Börner A, Badowiec A, Weidner S. Alterations in root proteome of salt-sensitive and tolerant barley lines under salt stress conditions. Journal of

Plant Physiology. 2015;**174**:166-176. DOI: 10.1016/j.jplph.2014.08.020

[119] Apel K, Hirt H. Reactive oxygen species: Metabolism, oxidative stress, and signal transduction. Annual Review of Plant Biology. 2004;**55**:373-399. DOI: 10.1146/annurev.arplant.55.031903.141701

[120] Hajheidari M, Eivazi A, Buchanan BB, Wong JH, Majidi I, Salekdeh GH. Proteomics uncovers a role for redox in drought tolerance in wheat. Journal of Proteome Research. 2007;**6**:1451-1460. DOI: 10.1021/pr060570j

[121] Hoekstra FA, Golovina EA, Buitink J. Mechanisms of plant desiccation tolerance. Trends in Plant Science. 2001;**6**:431-438. DOI: 10.1016/s1360-1385(01)02052-0

[122] Mahajan S, Tuteja N. Cold, salinity and drought stresses: An overview. Archives of Biochemistry and Biophysics. 2005;**444**:139-158. DOI: 10.1016/j.abb.2005.10.018

[123] Cornic G. Drought stress inhibits photosynthesis by decreasing stomatal aperture—Not by affecting ATP synthesis. Trends in Plant Science. 2000;**5**:187-188

[124] Osakabe K, Osakabe Y. Plant light stress. Encyclopaedia of Life Sciences. 2012. DOI: 10.1002/9780470015902.a0001319.pub2

[125] Chen TH, Murata N. Glycinebetaine: An effective protectant against abiotic stress in plants. Trends in Plant Science. 2008;**13**:499-505. DOI: 10.1016/j.tplants.2008.06.007

[126] de Souza CR, Maroco JP, dos Santos TP, Rodrigues ML, Lopes C, Pereira JS, et al. Control of stomatal aperture and carbon uptake by deficit irrigation in two grapevine cultivars. Agriculture, Ecosystems and Environment. 2005;**106**:261-274. DOI: 10.1016/j.agee.2004.10.014

[127] Maroco JP, Rodrigues ML, Lopes C, Chaves MM. Limitations to leaf photosynthesis in field-grown grapevine under drought—Metabolic and modelling approaches. Functional Plant Biology. 2002;**29**:451-459. DOI: 10.1071/PP01040

[128] Murata N, Takahashi S, Nishiyama Y, Allakhverdiev SI. Photoinhibition of photosystem II under environmental stress. Biochimica et Biophysica Acta (BBA)-Bioenergetics. 2007;**1767**:414-421. DOI: 10.1016/j.bbabio.2006.11.019

[129] Xue G-P, McIntyre CL, Glassop D, Shorter R. Use of expression analysis to dissect alterations in carbohydrate metabolism in wheat leaves during drought stress. Plant Molecular Biology. 2008;**67**:197-214. DOI: 10.1007/s11103-008-9311-y

[130] Boldt R, Edner C, Kolukisaoglu U, Hagemann M, Weckwerth W, Wienkoop S, et al. D-glycerate 3-kinase, the last unknown enzyme in the photorespiratory cycle in *Arabidopsis*, belongs to a novel kinase family. The Plant Cell. 2005;**17**:2413-2420. DOI: 10.1105/tpc.105.033993

[131] Araujo WL, Nunes-Nesi A, Nikoloski Z, Sweetlove LJ, Fernie AR. Metabolic control and regulation of the tricarboxylic acid cycle in photosynthetic and heterotrophic plant tissues. Plant, Cell & Environment. 2012;**35**:1-21. DOI: 10.1111/j.1365-3040.2011.02332.x

[132] Chmielewska K, Rodziewicz P, Swarcewicz B, Sawikowska A, Krajewski P, Marczak Ł, et al. Analysis of drought-induced proteomic and metabolomic changes in barley (*Hordeum vulgare* L.) leaves and roots unravels some aspects of biochemical mechanisms involved in drought tolerance. Frontiers

in Plant Science. 2016;7:1108. DOI: 10.3389/fpls.2016.01108

[133] Sweetlove LJ, Beard KF, Nunes-Nesi A, Fernie AR, Ratcliffe RG. Not just a circle: Flux modes in the plant TCA cycle. Trends in Plant Science. 2010;**15**:462-470. DOI: 10.1016/j.tplants.2010.05.006

[134] Kang G, Li G, Liu G, Xu W, Peng X, Wang C, et al. Exogenous salicylic acid enhances wheat drought tolerance by influence on the expression of genes related to ascorbate-glutathione cycle. Biologia Plantarum. 2013;**57**:718-724. DOI: 10.1007/s10535-013-0335-z

[135] Wei L, Wang L, Yang Y, Liu G, Wu Y, Guo T, et al. Abscisic acid increases leaf starch content of polyethylene glycol-treated wheat seedlings by temporally increasing transcripts of genes encoding starch synthesis enzymes. Acta Physiologiae Plantarum. 2015;**37**:1-6. DOI: 10 pp.206 ref.34

[136] Kosová K, Vítámvás P, Urban MO, Klíma M, Roy A, Prášil IT. Biological networks underlying abiotic stress tolerance in temperate crops—A proteomic perspective. International Journal of Molecular Sciences. 2015;**16**:20913-20942. DOI: 10.3390/ijms160920913

[137] Gratani L, Pesoli P, Crescente M, Aichner K, Larcher W. Photosynthesis as a temperature indicator in *Quercus ilex* L. Global and Planetary Change. 2000;**24**:153-163. DOI: 10.1016/s0921-8181(99)00061-2

[138] Keller M, Bokszczanin KL, Bostan H, Bovy A, Chaturvedi P, Chen Y, et al. The coupling of transcriptome and proteome adaptation during development and heat stress response of tomato pollen. BMC Genomics. 2018;**19**:447. DOI: 10.1186/s12864-018-4824-5

[139] Wahid A, Gelani S, Ashraf M, Foolad MR. Heat tolerance in plants: An overview. Environmental and Experimental Botany. 2007;**61**:199-223. DOI: 10.1016/j.envexpbot.2007.05.011

[140] Arce D, Spetale F, Krsticevic F, Cacchiarelli P, Las Rivas JD, Ponce S, et al. Regulatory motifs found in the small heat shock protein (sHSP) gene family in tomato. BMC Genomics. 2018;**19**:860. DOI: 10.1186/s12864-018-5190-z

[141] Sung DY, Kaplan F, Guy CL. Plant Hsp70 molecular chaperones: Protein structure, gene family, expression and function. Physiologia Plantarum. 2001;**113**:443-451. DOI: 10.1034/j.1399-3054.2001.1130402.x

[142] Chinnusamy V, Zhu J, Zhu J-K. Cold stress regulation of gene expression in plants. Trends in Plant Science. 2007;**12**:444-451. DOI: 10.1016/j.tplants.2007.07.002

Chapter 10

Seed Priming: The Way Forward to Mitigate Abiotic Stress in Crops

Melekote Nagabhushan Arun, Shibara Shankara Hebbar, Bhanuprakash, Thulasiram Senthivel, Anil Kumar Nair, Guntupalli Padmavathi, Pratima Pandey and Aarti Singh

Abstract

Abiotic stress is a major threat to the farming community, biasing the crop productivity in arid and semi-arid regions of the world. The seed is an important component of agriculture, contributing significantly to the booming production of food and feed crops across the different agro-ecological regions of the world with constant challenges with reference to production, storage, and quality control. Germination, plant growth, and development *via* non-normal physiological processes are detrimentally affected by stress. Seed priming is an alternative, low cost, and feasible technique, which can improve various abiotic stress tolerances through enhanced and advanced seed production. Seed priming is a process that involves imbibing seed with a restricted amount of water to allow sufficient hydration and advancement of metabolic processes but preventing germination. The beneficial influence of priming on the germination performance of diverse species is attributed to the induction of biochemical mechanisms of cell repair: the resumption of metabolic activity that can re-impose cellular integrity, through the synthesis of nucleic acids (DNA and RNA) and proteins and the improvement of the antioxidant defense system metabolic damage incurred by dry seed and thus fortifying the metabolic machinery of the seed. With this background, this chapter highlights the morphological, physiological, biochemical, and molecular responses of seed priming and recent advances in priming methods as a tool to combat abiotic stress in crop plants.

Keywords: abiotic stress, seed priming, crop establishment, physiological changes, biochemical changes

1. Introduction

Crop plants are subjected to multiple abiotic stresses during their life span that significantly reduces productivity and foreshadow global food security. The improvement of crop plants through direct selection-based conventional breeding for drought and salinity stress appears to be quite complex [1]. Abiotic stresses such as salinity, drought, flooding, heat, cold, freezing, excess light, UV radiation, and heavy metal toxicity have a significant impact on seed which reduce germination rate, seedling growth and yield with significant variations from crop to crop worldwide [2]. Water stress is a key agronomic problem globally and is one of the most important factors reducing crop productivity. Drought is one of the most important environmental component limiting plant growth and productivity. The advancement of crop plants through direct selection-based conventional breeding for drought and salinity stress appears to be quite complex. In agriculture, alternative innovative technologies such as plant tissue culture, seed priming and genetic engineering could play a major role in combating abiotic stresses for increasing yield. The quality of seed in the present time can be accomplished by various means where the basic and applied knowledge of plant physiology, genetics and seed technology all are integrated to improve the criteria of quality of seeds. Seed germination is a dynamic and complex stage of plant ontogeny, with several interactive metabolic processes changing from a storage phase to a mobilization phase [3]. The multitude of simultaneous metabolic processes taking place in germinating seed makes it difficult to explorte events related to the initiation of the germination process. The catabolic processes in the storage tissues can confound measurements and interpretation of anabolic and growth processes occurring in the developing embryo in the entire seed. The time from sowing to the seedling establishment is of considerable importance in crop production and has major impacts on plant growth, final yield and post-harvest seed quality [4]. Efficient seed germination is important for agriculture. Successful foundation of early seedling indeed requires an accelerated and uniform emergence and root growth. In nature, plants often face the challenge of severe environmental conditions, which include various biotic and abiotic stresses that exert unfavorable effects on plant growth and development causing considerable losses in crop productivity. Drought, salinity and extreme temperature cause osmotic stress on crop plants which generate an imbalance at cellular, molecular and physiological levels which ultimately lead to plant death [5]. Abiotic stress causes many physiological and biochemical changes in the seedlings, which include the generation of reactive oxygen species (ROS), leading to membrane injury and cell leakage and destruction of photosynthetic components [6]. Seed priming is a simple, safe, cost effective, and effective approach for enhancement of seed germination, early seedling growth, and yield under stressed and non-stressed conditions [7]. Seed priming is a form of seed preparation in which seeds are pre-soaked before planting with a certain solution that allows partial hydration but will not germinate and redried to original moisture content [8].

2. Abiotic stress and its effect on crop growth and development

Abiotic stresses are closely related individually or cohesive; they cause physical, morphological, biochemical and cellular changes that adversely affect plant growth and productivity and ultimately yield. Heat, drought, salinity and cold are major abiotic pressures that cause severe cell damage to a wide variety of plants, including crop plants (**Figure 1**). Water stress or drought is a major challenge for agricultural production worldwide. Excessive wilting causes a change in the ratio of membrane lipids and this may result in increased electrolyte leakage [9]. Drought is known as a prime abiotic factor that limits plant growth and production in arid and semi-arid regions and is the most significant factor in world security and sustainability in agricultural production. Drought slows growth, induces stomatal closure and therefore reduces photosynthesis, growth and yield in a number of plant species [10]. Water stress has been revealed to be one of the factors limiting the productivity of cowpea as it affects not only the production of the grain, but also the whole process of growth of all organs of the plant and its metabolism [11]. Water stress limits the size of individual leaves and leaf number. Physiological disorders occur during seed germination under abiotic stress is a decrease in water uptake by the seed due to the low water potential of the germination medium. Slow or abnormal growth

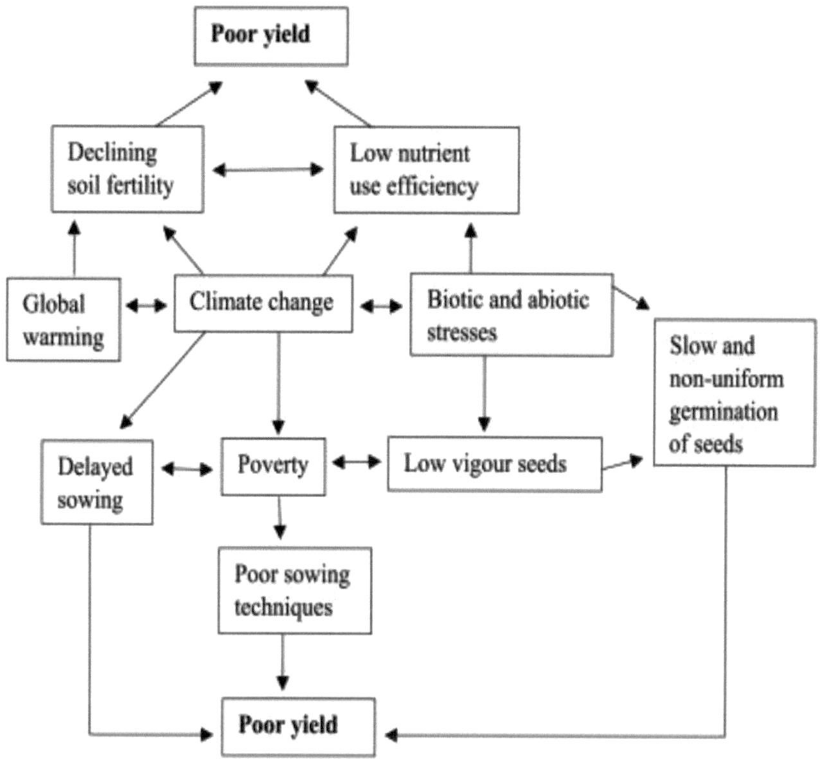

Figure 1.
Direct and indirect causes of low productivity of field and horticultural crops.

and emergence result in fewer and smaller plants, which are more susceptible to various abiotic stresses [12]. In addition to causing various structural changes at different levels of organization in the seed, slow rate of imbibitions may lead to a series of metabolic changes, including up regulation or down-regulation of enzyme activities, perturbance in the mobility of inorganic nutrients to developing tissues, disturbance in N metabolism, imbalances in the levels of plant growth regulators, reduction in hydrolysis and utilization of food reserves and accumulation of compatible osmotic such as soluble sugars, free proline and soluble proteins [13]. Stress processes may lead to poor or complete lack of germination under abiotic conditions. Salinity is also considered as substantial abiotic stress and significant factor affecting crop production globally and especially in arid and semi-arid regions [14]. The salinity of soil reduces water availability of plant root via negative (low) osmotic potential as well as decrease of germination dynamics of plant seeds by ionic toxicity of Na^+ and Cl^- [15]. Of the extensive forms of abiotic stress, heat stress has an independent mode of action on the physiology and metabolism of plant cells. Due to high temperatures, various physiological injuries have been observed such as scorching of leaves and stems, leaf abscission and senescence and root and shoot growth inhibition or fruit damage, which as a result lead to decreased plant productivity [16]. The plant growth is reduced by affecting the shoot, net assimilation rates and finally the total dry weight of the plant due to high temperature [17]. The stress is extensively prominent on reproductive development than on vegetative growth, and the sudden reduction in yield with temperature is mainly associated with pollen infertility in many crop species under effects of high-temperature [18]. Heat stress which causes multifarious, and often adverse, alterations in plant growth, development, physiological processes, and yield is one of the major consequences of high-temperature stress [19]. Due to stress limitation of available technology, we should think of various alternate technologies such as priming, mutagenesis, and tissue culture for combating abiotic stresses. In crop species, seed germination and early development of seedling are the most sensitive stages to salinity stress. Salinity delays the onset, reduce the rate and increase the dispersion of germination phenology, leading to a reduction in plant growth and final crop yield. Thus, if the effect of abiotic stress can be mitigated at the early stages, the chance of establishing a successful crop under stress will be improved [20].

3. Seed priming and its importance

Priming involves prior exposure to elicitors which brings a cellular state that arrest the harmful effects of abiotic stress, and plants raised after priming are more tolerant of abiotic stress. Seed dormancy is an emerging issue related to germination that is common to many plant species. It is a practice that allows a species to estimate the germination period in a population. Some species use a environmental indicator (such as drought, rain or temperatures) to integrate germination of many seeds at a particular time of year. Temperature, humidity, air and light conditions are the main factors for seed germination. The minimum temperature is the minimal temperature at which a seed germinates effectively and the maximum is the highest temperature at which seeds can germinate [21]. At optimal temperatures, germination is rapid and uniform but with meager change in this temperature can damage seeds or make them go into the dormancy condition. Seeds need accurate moisture to initiate internal processes leading up to germination. Osmotic adjustment or priming of seeds before sowing is known as a potent way to increase germination and emergence rate in some species with incremental seed development [22]. Inadequate seed germination and subsequently poor field establishment are a common occurrence at adverse conditions of the environment. Seed germination and early seedling growth are the most sensitive stages of water limitation and the water deficit may impede the onset and reduce the rate and uniformity of germination, leading to poor crop accomplishment and yield in several crops. Seed priming is a water-based approach with low external water potential that restrict hydration (controlled hydration of seed) and permit metabolic processes necessary for enhancing germination rate and seed quality by managing the seed moisture content and temperature in which the seed is taken through the initial biochemical processes within the initial stages of germination but preventing the seed transition towards full germination [23]. This will assure better field emergence and disease resistance under various adverse conditions. The purpose of priming is to reduce the germination time and improve stand and germination percentage under unfavorable environmental conditions. Primed seeds are used instantly but may be dried and stored for a short time for later use. Primed seeds attain the potential to rapidly imbibe and revive seed metabolism thus enhancing the germination rate [24]. These attributes have practical agronomic implications notably under adverse germination conditions. Pre-treatment using a priming induced stimulus like sodium nitroprusside, hydrogen peroxide, melatonin and polyamines results in enhanced cell tolerance and amelioration of stress-induced plant growth inhibition [25]. Therefore, the beneficial effects of priming may be more evident in under favorable rather than unfavorable conditions [26]. Primed seeds generally exhibit an increase germination rate, greater germination uniformity, and, at times, greater total germination percentage. Abiotic stresses produce oxidative stress and activate similar cell signaling pathways and cellular responses [27]. But, seed priming seems to activate these signaling pathways during the early stages of growth, resulting in faster defense responses [28]. The abiotic stress tolerance generated by seed priming is accorded via the synchronization of divergent physiological, biochemical, systemic, cellular and molecular modulations [29]. The purpose of seed germination is to reduce the germination period and to protect the seed from environmental stress during the critical stage of seedling growth to integrate the growth leading to uniform establishment and improved yields. It minimizes the effect of salinity on the morphological parameter of the plants. One of the priming methods called osmopriming is a commercially available method of improving seed germination and strength. It controls seed imbibition to initiate the germination process followed by the seed drying up to its original weight. Various methods of seed planting such as hormonal priming and support for chemo priming in improving the process of germination, flowering and growth of plants are done for commercial purposes on the farm or on the farm. Planting seed on a farm requires electricity, high-tech seed harvesting, seed hardening or seed-drying process is accessible to farmers to help them with cool farming and horticulture [30]. The effective use of a seed management method depends on the type of test, method of application, crop selection, early performance of the plant, chemical selection, concentration, duration of treatment and the purpose of the application. Priming method in seed management techniques is proven very essential factor for enhancing quality issues, germination rate and establishment. Priming can interfere with some of the degenerative effects of aging, leading to improved seed performance [31]. It has shown an immense effect of priming to activate different processes related to cell cycle and to induce to the synthesis of nuclear DNA in radial tip cells [32]. Seed priming had the most beneficial effect on leaf area indicating the advantage of rapid seedling emergence [33]. The reason attributed may be due to cell division, cell number due to multiplication in various plant tissues, auxin multiplication, cell wall plasticity and permeability of cell membrane, increases photosynthesis, cell enlargement and rapid cell wall elongation [34].

4. Factors affecting seed priming process

Seed priming is controlled by many factors such as aeration, light, temperature, concentration of priming solution, time (duration), and seed quality.

4.1 Aeration

Aeration is considered an important step to assist seed respiration, seed viability and contributes to synchronize the emergence [35] Germinating seeds respire very actively and need sufficient oxygen. The consequence of aeration varies according to species: in onion aeration of the PEG solution increased the germination process compared to non-aerated treatment [36]. No difference was noticed in the germination of lettuce seeds between aerated and non-aerated K_3PO_4 priming [37].

4.2 Duration

Duration (maximum length of time) of priming is one of the key factors for seed priming. Seed priming for 7 h and 14 h is sufficient to augment seed and seedling vigor, stand establishment, and grain yield contrary to soaking the seeds for 21 h for optimizing duration of hydro priming in green gram [38]. Seeds primed for 12 h took significantly fewer days to emerge and reach maturity when compared to the untreated dry seed, whereas 36 h primed seeds showed poor germination and 48 h primed seeds inhibited germination. This inhibition may be attributed to the prolonged period of priming that led to excess water in the seeds and greater reduction in the O_2 availability to the embryo [39].

4.3 Temperature

The lower temperatures during priming seem to result in slower imbibition of seeds, increasing the duration of phase II of the triphasic pattern of imbibitions [40]. This would allow the activation mechanisms to repair the membrane systems and prevent destruction caused by rapid imbibition. Hydro priming at 15°C increased synchronization of germination and speed of seedling emergence in *Solanum lycocarpum* [41]. Seed priming at 15°C showed good vigor, mean germination time and growth response in french bean compared to seed primed at 30°C which inhibits germination [42]. Seed priming at low temperature showed the beneficial effects on germination index, speed of germination, length of radical and plumule, fresh weight of seedling and seedling vigor index compared to high temperature in wheat [43]. Priming was effective in reducing the time for 50% germination, mean germination time and increase germination percent and seedling vigor index when primed with potassium di-hydrogen phosphate at low temperature in sunflower [44]. GA_3 (20 ppm) primed seeds at low temperature showed significantly higher germination, root length, shoot length and seedling dry weight over higher temperature in soybean [45]. Seed priming at 15°C showed better percent increase in germination, mean germination time than at higher temperature at 30°C. Seed priming at low temperature associated with a buildup of nucleic acids and protein synthesis and membrane repair [46].

4.4 Concentration

Seedling growth was not proper and subsequent growth of seedling was arrested probably due damage of cell organelles due to higher concentration when seed primed with ethrel. Increasing the concentration of priming solution from 10^{-3} M to 10^{-1} M at constant temperature decreased the germination to the extent of 21% and 56% when papaya seeds were treated with oxalic acid and mannitol [47]. Soaking of wheat seeds in GA_3 at low concentration not only enhances the speed of germination but also increase the length of radical as well as plumule [48]. Osmopriming with PEG at low concentration in spinach improved the final germination percentage, germination rate and uniformity [49]. Seed priming at lower concentration of growth regulators favors the increase enzymatic activity which leads to the favorable environment for the germination. Growth regulators at higher concentration inhibit the seed germination which might have been due to solute leakage and lipid per oxidation which limit the necessary material for germination and seedling growth [50]. Tomato seeds osmo-primed with PEG 6000 at low concentration improved mean germination time, seedling emergence percentage and cell membrane stability over higher concentration [51]. Seed priming with 1% sodium molybdate reduced the seed germination due to toxic effect on physiological and biochemical processes within the cell [52]. In case of seed priming with higher concentration of ammonium molybdate and magnesium nitrate solution germination was absent. The reason attributed due to higher concentration of chemical which cause detrimental effect on cellular mechanism and mitochondrial; membrane in seed [46].

4.5 Light

Light effect changed widely according to crop species. Illumination during seed priming of celery seeds reduced dormancy [53]. The best germination in lettuce was obtained with seeds primed in the dark [37]. Light played a vital role in maximizing seed germination with guava seeds primed at 12 h of light and 12 h of dark [54].

5. Seed priming and physiological changes during priming

Seed treatment technology is an important link between seed producers and crop production. Seed priming is the technology which is a novel concept of treating seeds using various solvents including water which activates physiological processes of seeds. When dry seeds are immersed in water, water absorption occurs in three stages [55]. Stage I is imbibition where there is a swift initial water uptake due to the seed's low water potential. Proteins are synthesized using existing mRNA and DNA and mitochondria are repaired during stage I phase. In stage II, there is a steady increase in seed water content, but physiological activities associated with germination are initiated, including synthesis of proteins by translation of new mRNAs and new mitochondria. There is a swift uptake of water in stage I where the process of germination is completed culminating in radicle emergence Stages I and II are the foundations of successful seed priming where the seed is brought to a seed moisture content that is just short of radicle protrusion [56]. The pattern of water uptake during priming is identical to that during slow germination and controlled. Seed hydration triggers germination via three stages: imbibition, lag phase and radicle protrusion through the testa [57]. Seed requires oxygen, water, and a suitable temperature for germination. The time from sowing to the seedling establishment is of considerable importance in crop production and has a major impact on plant growth, yield and post-harvest seed quality [4]. During germination process of orthodox seeds three distinct phases is manifested where in (1) Phase I: seed hydration process related to passive imbibition of dry tissues associated with water movement preliminary occurring in the apoplastic spaces; (2) Phase II: activation phase associated with the rejuvenate of metabolic activities and repairing processes at the cell level; and (3) Phase III: initiation of developing processes associated to cell elongation and leading to radical protrusion. Phases I and III both entail an increase in the water content while hydration remains stable during Phase II. Before the conclusion of Phase II, it is considered that germination remains a reversible process: the seeds may be dried again and remain functioning during storage and able to subsequently re-initiate germination under propitious condition. Water-based seed priming is elucidated as a pre-sowing dressing that partially hydrates seeds without allowing emergence [58]. Different treatments may indeed be applied during the reversible stage of germination (point 3). They broadly differ according to the osmotic potential of the priming solution, the duration, the external temperature, and the existence of specific chemical compounds. The efficient treatments trigger metabolic processes activated during phase II of germination, which are then transitory stopped before a loss of desiccation occurs [59]. Priming is a technique that allows controlled seed hydration to trigger pre-germinative metabolism but does not allow the seed for the transition towards full germination. In the case of primed seed hydration treatment allows regulated imbibition and induction of the pre-germinative metabolism ("activation"), but radicle emergence is prevented, represented by the extended second phase. Final phase (phase III) represents the germination and post-germination phase which is again similar in the case of non-primed seeds. In the case of primed hydration treatment seeds allow regulated imbibition and the introduction of pre-germinative metabolism ("initiation"), but the emergence of radicle is inhibited, represented by an extended second phase. The final stage (phase III) represents germination and the post-emergence phase which is similar in the case of unprimed seeds. During Phase I Imbibition there is a rapid absorption of water due to the lower seed potential than outside. Initially there is water movement in apoplastic spaces, proteins are synthesized from existing mRNAs and DNA and mitochondria are repaired during Phase I. In phase II, there is the activation of metabolic and repairing activity along with the synthesis of proteins by translation of new mRNAs of new mitochondria, where as phase III is associated with regaining capacity of rapid water uptake and initiation of growing processes linked with cell elongation that leads to radical protrusion. Priming allows a seed to hydrate up till a seed moisture content involving the entire phase I and before the end of phase II when the germination remains a reversible process just short of the radical protrusion [56]. Thus priming activates 'pre-germinative metabolism' that included a wide range of physiological functioning. This activates DNA repair pathways, ROS scavenging systems (that impart for seed repair response) and also helps in preserving genome integrity [60]. Priming solutions can be supplemented with hormones or beneficial micro-organisms. The seed may be dried back for storage, distribution and planting. Priming can induce the germination by improving the speed and synchronization of seed germination [61]; it can improve seed vigor which require very short or no activation time during germination. The advanced germination status of primed seeds contributes to increased germination under stressful conditions [62]. Besides it also facilitates the initiation of many germination-related activities such as enhanced energy metabolism, early reserve mobilization, embryo expansion and endosperm weakening [31]. Priming also enhances the specific stress-responsive systems which include induced accumulation of LEA and heat shock proteins [63], activation of catalase and other antioxidant scavenging enzymes [64] and up-regulation of genes encoding peroxiredoxin [65].

6. Biochemical and molecular basis of stress tolerance

The type of test, method of application, selection of crop, initial performance of the crop, selection of chemical, duration of treatment, its concentration, and the purpose of implication helps in successful application of seed management technique. Priming method in seed management techniques is established, which is very important factor for enhancing quality issues, germination rate and establishment. Priming can improve some of the aging-induced deteriorative events, resulting in improved seed performance [31]. It has shown an immense effect to activate different processes related to cell cycle and to induce synthesis of nuclear DNA in radial tip cells in tomato [32]. Prolonged storage of seeds resulted in a decrease in protein content which led to an increase in oxidation of amino acids, due to increased respiratory function and progression in the process of deteriorating stored seed. Seed degradation results in loss of membrane integrity, changes in enzymatic functions and reduction of protein and nucleic acid synthesis and lesions in DNA [66]. Priming with NaCl and 30% PEG for 24 h of rice seed initiated in increase in the activity of superoxide dismutase (SOD) and peroxidise (POD) which enhance the intensity of respiration of plant and cause an increase in vigourity in germination [67]. Priming is also thought to increase the activity of many enzymes involved in metabolism of carbohydrates (α- and β-amylases), proteins (proteases) and lipids mobilization (iso citratelyase) that are implicated in the stored reserves mobilization [68]. These enzymes are vital in the breakdown of macromolecules for the development and growth of the embryo that ultimately result in early and higher seedling emergence [69]. There are reports that priming facilitates the repair of chromosomal damage [70], permits early DNA replication and repair, increases RNA and de novo protein synthesis and reduces the leakage of metabolites [24]. Thus, total seed protein, peroxidases, polyphenol oxidases, RNA and de novo protein synthesis were enhanced significantly by seed priming. Among the various processes of priming, osmopriming could enhance rapid seed germination by reducing mechanical hindrance on the germinating embryo. The pre-treatment of seeds with priming agents facilitates the active absorption of ionic molecules with greater ATP availability and repair of deteriorated seed parts for reducing leakage of metabolites leading to faster embryo development [71]. It also, reflected in greater cellular membrane integrity, counteraction of lipid per oxidation, and free radical chain reaction often are found to be directly correlated with the maintenance of viability and reduce moisture uptake by hydrated-dehydrated seed [72], repair of biochemical lesions by the cellular enzymatic repair system [73] and metabolic removal of toxic substances [74], counteraction of free radical and lipid peroxidation reactions [75], biochemical changes like enzyme activation [76], and improvement of germination rate particularly in old seeds [77]. Priming provides a 'head-start' of seed transition from quiescent to germinating state, thus increasing the potential to germinate. Seed priming thrust abiotic stress on seeds that represses radicle protrusion but stimulates stress-responsive elements [78].

7. Reversal of seed deterioration by priming

Seed deterioration is defined as the loss of seed viability and vigor due to aging effects and adverse environmental factors distinctly higher temperature, relative air humidity and oxygen/carbon-dioxide ratio [66]. Seed deterioration is associated with several cellular, metabolic and chemical alterations including lipid per oxidation, membrane disruption, and DNA damage, impairment of RNA and protein synthesis and causes several detrimental effects on seeds [79]. The cause of seed deterioration is damage to cellular membranes and other sub cellular components by harmful free radicals generated by peroxidation of unsaturated and polyunsaturated membrane fatty acids. Seed storage causes a decrease in the protein content which may be related to oxidation of the amino acids due to the increase in the respiratory activity and advance in the deterioration process of the stored seeds [80]. Poor storage conditions may accelerate seed deterioration of seeds [81]. As seed deterioration increases, seed performance progressively decreases. Plants that originated from deteriorated seed can reduce growth rate. The aging of seeds, during long term storage deteriorated their vital status which was expressed in change in their moisture content, decreasing of their sowing qualities and development of weaker seedlings with higher water content [82]. The main mechanism for aging of seed is associated with increased peroxidation of lipid membranes [65]. Priming can reverse some of the aging induced deteriorative factors and thus improve seed performance [31]. The beneficial effects of priming are associated with the repair and building up of nucleic acid, increased synthesis of proteins as well as the repair of both mitochondria and membranes [24]. Priming for 24 h with GA_3 and ammonium molybdate in aged seeds showed increase enzyme activity restored almost entire protein profile and esterase and peroxidase isozyme profile as it allowed repair system to combat sub-cellular damage and activated synthesis of enzymes and protein [83]. Under invigorated, metabolic repair processes in deteriorated seeds occur before onset of seed germination process [84]. Seed priming is more useful for enhancing germination of low-quality seed lots than higher-quality ones which indicates that repair of aging is one of the primary advantages [85]. Significant changes in enzyme activities were observed in primed seeds compared to un-primed seeds. Desiccation and storage of seeds has been suggested to result in progressive loss of integrity of the membrane components of the seeds, which in turn bring about to seed deterioration as measured by loss of seed vigor and viability. Maintenance of the integrity of DNA by repairing the damages incurred naturally is crucial for generating error free template for transcription and replication with fidelity. The damage to DNA which accumulates during the seed aging is repaired by aerated hydration [86]. During imbibition prior to germination the integrity of cell membranes need to be re-established. Rapid imbibitions by the seed at this time probably reverse the damage and cell will attain maximum vigor by repair mechanism. It is thought that hydro priming initiates an oxidative stress, which generates reactive oxygen species, and catalase is synthesized to minimize cell damage. In addition to catalase, levels of superoxide dismutase, another essential enzyme quenching free radicals also increases during priming. Increased levels of these free radical scavenging enzymes due to the oxidative stress during priming might also protect the cell against membrane damage due to lipid peroxidation occurring naturally [68]. Priming with GA_3 and ammonium molybdate allowed repair system to combat sub-cellular damage activated enzyme synthesis due to accelerated aging. The changes in the activation of the enzymes, upon priming advocate that mobilization of storage material may be responsible for increased germination and vigor in primed seeds when compared to unprimed aged seeds [87].

8. Effect of priming on reserve mobilization and management of oxidative status

It is proposed that germination-related processes such as respiration, energy metabolism, and initial reserve mobilization also occur during priming. Higher respiratory activity is required to cover energy pool for speed up germination. Increased respiratory activity has been associated with pre-sowing treatments. Seed priming increased the respiratory activity of seeds and reduced the oxygen-time constant and increased the standard deviation of germination responses [88]. During seed germination, storage proteins, which provide a source of reduced nitrogen and inorganic minerals, need to be mobilized to support seedling growth [89]. Soluble protein content increased in pepper seeds after 12 days of priming in –1.34 MPa NaCl solution [90]. Pepper seedlings developed from primed seed had improved soluble protein [90]. Osmo-priming induced accumulation of stress proteins, such as late embryogenesis abundant (LEA) proteins and heat shock proteins (HSP) [91]. Management of oxidative status is also an important part of primed seed physiology. Priming activates the response of the antioxidant system and modifies the prepared seeds for potential stresses [92]. In the early stages of seed intake and germination, the production of reactive oxygen species (ROS) is primarily due to the respiratory activity of mitochondria, β-oxidative pathway activity, and enzymes such as NADPH oxidase, extra-cellular peroxidase, and oxalate oxidase [93]. Antioxidants, by breaking down high ROS during early endocytosis, play an important role in ensuring successful germination, especially under stressful conditions [94]. Seed priming in tomato seeds revealed enhanced activity of antioxidant enzymes such as ascorbate peroxidase, catalase, peroxidase, glutathione reductase and superoxide dismutase [95]. Free radical scavenging enzymes such as catalase and super dismutase are synthesized during hydro priming to defend the cell from damage due to lipid peroxidation, which occurs due to the oxidative stress induced by hydro priming. Priming synchronizes all the cells of the germinating embryo in the G2 phase of the cell cycle so that upon further imbibitions, cell division proceeds uniformly in all the cells ensuring uniform development of all parts of the seedlings. Seed priming of pepper (*Capsicum annuum* L.) conducted under temperature stress (low 15° C and high 35°C) for two consecutive runs revealed enhanced germination even in stressful conditions. Priming was found to enhance repair of membranes, the activities of hydrolytic enzymes, and antioxidant system. However, it was noticed that priming decreased sucrose content, whereas the fatty acid composition remained unchanged and increased enzymatic activity of catalase which was enhanced significantly in pepper seeds [96]. Osmopriming with –1.5 MPa PEG 6000 for 6 days of aged seeds of sweet pepper resulted in an improved germination with decreased levels of malondialdehyde (MDA)and total antioxidant activity, total ascorbate, de-hydro ascorbate, and catalase activity in primed seeds enhanced the defense mechanism in protecting the cell membrane damage from reactive oxygen species [29]. Nano priming augmented the performance of seeds by enhancing α-amylase activity, increasing soluble sugar content to support early seedling growth, up-regulating the expression of aquaporin gene in germinating seeds, increased stress tolerance through lower ROS production and creation of nano pores for enhancing water uptake in crops in field [74]. The main obstacle to the practical application of primed seeds is storage and viability. This barrier can be overcome by knowing the genes/markers associated with seed germination and the identified markers can be used to assess the effect of priming on germination efficiency and seed vigor [97]. Genes/markers involved in rice seed priming were identified by comparing differential proteins between the dry and imbibed seed using two-dimensional electrophoresis [98]. Improved germination was reported in polyethylene glycol-6000 (PEG-6000) primed rapeseeds which, on germination, resulted in differential expression of 952 genes and 75 proteins [99].

9. Methods of seed priming and role in improving crop productivity

Seed priming technique such as hydro priming, halo priming, chemical priming, osmopriming, hormone priming, solid matrix priming and nutrient priming are extensively used in crop plants for many environmental stresses. Seed priming increases germination and growth especially under environmental constraints. However, the degree of efficacy of different priming agents varies with plant species and diverse environmental conditions [100]. Different seed priming methods employed to mitigate stress and salt tolerance as reported by many researchers are shown in **Tables 1–5**.

S. no	Priming agent	Crop	Attributed improved	References
1.	Water (12 h)	Rice	Accelerated germination, early emergence time, increased initial growth after emergence, increased dry root weight and dry matter productivity	[101]
		Wheat	Higher germination percent, increased water use efficiency, homogeneity of seedling emergence, increase growth and yield parameters. Aged seeds when primed with water improved induced increase in enzyme activity, improved germination and seedling characteristics	[102]
		Pearl millet	Increased crop emergence and crop yield in arid zone	[103]
		Chickpea	Improved membrane integrity and electrical conductivity of seed leachate. Increased germination indices and seedling growth	[104]
		Sunflower	Increased germination percentage, seed vigor index and seedling growth rate. Reduced time to 50% seedling emergence	[105]

Table 1.
Hydropriming technique and their effectiveness in improving growth of various crops under adverse condition.

S. no	Priming agent	Crop	Concentration	Attributed improved	References
1.	PEG	wheat	−1.0 MPa	Seed germination and grain yield	[106]
		Rape seed	−0.5 MPa	Percentage of germination under saline soil, seedling length, and increased metabolic process in seeds	[107]
		Chickpea	−0.5 MPa	Seed germination, yield and improvement in seed quality attributes	[108]
2.	Mannitol	Broad bean	1%	Percentage of germination, higher seedling length and seed vigor index	[109]
		Rice	1%	Enhanced germination percentage, growth of seedlings and plant survival under salt stress	[110]
3.	Sorbitol	Wheat	1%	Improved seed vigor, plant morphology and upregulation of plant growth regulator	[111]

Table 2.
Osmoconditioning techniques and their effectiveness in improving growth of various crops under adverse condition.

S. no	Priming agent	Crop	Attributed improved	References
1.	Gibberellic acid (50 ppm)	Rice	Improve crop emergence, crop establishment and yield in direct seeded rice	[112]
2.	Cytokinins (100 ppm)	Soybean	Hasten seed germination rate and seedling development. Improving root length and enhanced nutrient uptake and water use efficiency	[113]
3.	IAA (20 mg L^{-1})	Cotton	Improves the germination, root length, seedling height and seedling growth, biomass and leaf photosynthesis capacity and yield	[114]
4.	Auxin (50 ppm)	Wheat	Increased grain filling rate and grain yield and positive effect of photosynthesis	[115]
5.	Salicyclic acid (0.9%)	Sesame	Increased germination percent and seedling length and vigorous growth, reduced germination time	[116]

Table 3.
Priming with plant growth regulators and their effectiveness in improving growth of various crops under adverse condition.

9.1 Hydropriming

Soaking seeds with water overnight and then drying before sowing markedly improved seedling emergence, plant growth establishment, vigor and final yield in field crops [12]. Slow and inconsistent germination of seeds has prompted the need for water-based seed priming. Hydro priming is a very sustainable, cost-effective and environmentally friendly technique, mainly involving soaking the seeds in water for a predetermined time and then drying them back to their initial moisture level [89]. The process of seed germination occurs in three phases, viz., rapid water uptake or imbibition (phase I), lag or plateau phase (phase II), and protrusion of seminal root and resumption of growth (phase III) [55]. Hydro priming reduces the

S. no	Priming agent	Crop	Attributed improved	References
1.	Potassium nitrate (2%) (50 ppm)	Rice	Improve crop emergence, crop establishment and yield in DSR	[117]
2.	$CaCl_2$(50 mM)	Sorghum	Increased germination rate, root and shoot length under salt stress condition	[112]
3.	KH_2PO_4 (1%)	Maize	Increased field emergence, plant height number of leaves and seed yield	[26]
4.	$Mg(NO_3)_2$ + $ZnSO_4$ (2%)	Wheat	Increased plant height, number of leaves, leaf area and chlorophyll content and increased yield under drought stress	[118]
5.	Ammonium molybdate (0.1%)	Common Bean	Improved germination percent, net CO_2 assimilation rate, chlorophyll content and increased grain yield	[119]
6.	KCl (1%)	Green gram	Enhancing crop stand and increasing yield under drought condition	[120]

Table 4.
Nutrient priming techniques and their effectiveness in improving growth of various crops under adverse condition.

S. no	Priming agent	Crop	Attributed improved	References
1.	*Rhizobium sp* + *Trichoderma viride*	Green gram	Increased germination percentage, synchronized seed germination, growth and yield components.	[121]
2.	*Bacillus amyloliquefacians*	Rice	Enhanced activities of peroxidase and polyphenol oxidase in seedlings. Improved germination percentage, increased leaf area	[122]
3.	*Bacillus* sp. (MGW9)	Maize	Improved the germination energy, seedling length, relative water content, field seedling emergence and seedling growth	[123]
4.	*Trichoderma asperelllum* (24 h)	Wheat	Plant growth promoting activities, uniformity in seed emergence, good seedling vigor and establishment under stress conditions	[124]
5.	*Azotobacter chroococuum*	Chickpea	Implant plant growth, and dry weight and yield	[125]
6.	*Azospirillum lipoferum*	Barley	Tolerance to stress and improved plant growth and productivity	[126]

Table 5.
Biopriming priming techniques and their effectiveness in improving growth of various crops under adverse condition.

lag period ensures rapid and uniform germination for good stand establishment [127]. Controlled seed hydration as a pre-sowing strategy triggers pre-germination metabolic activities in the form of cellular physiological, biochemical, and molecular changes [93]. Ameliorated germination of hydro primed seeds is a repercussion of stimulation of enzymes (amylase, protease, phosphatase, lipase, etc.), ATP production, RNA and protein synthesis, DNA replication, detoxification of ROS and lowering of lipid per oxidation by antioxidant enzymes [superoxide dismutase (SOD), catalase (CAT), peroxidase (POD), and glutathione reductase (GPx)], accumulation of germination enhancing metabolites (proline, soluble sugars, etc.), higher utilization of seed reserves (proteins, carbohydrates, lipids, and phosphorus compounds), and supplementary metabolic repairing mechanisms. The major limitation associated with hydro priming is uncontrolled water uptake which result in unsynchronized germination [12]. Plants produced from hydro-primed seeds had substantial water uptake which is positively associated with seedling growth. Hydro-priming is a simple method to improve abiotic stress tolerance and improve germination percentage in cauliflower [28]. Seed soaking in water improved germination, seedling establishment and yield in wheat crop under contrasting environmental conditions [128]. Hydro-priming of rice seeds improved the germination rate, speed and uniformity even under less than optimum field condition in upland conditions [129]. Hydropriming of soybean seeds for 12 h was effective to increase number of pods, grain yield and biological yield under drought stress conditions [130]. Hydro-priming for 48 h of naturally aged rice seeds were more efficient in seed germination, emergence and seeding vigor under direct seeded rice [131].

9.2 Osmoconditioning

Osmopriming is known as osmotic priming, osmotic conditioning or osmoconditioning. It is a pre-sowing treatment in an osmotic solution that allows seeds to imbibe water to proceed to the first stage of germination but prevent radicle protrusion through the seed coat [132]. In this methodology, seeds are soaked in osmotic solutions of organic compounds such as polyethylene glycol, mannitol, glycerol, and sorbitol having low water potential so as to regulate the water uptake by seeds and allow the pre-germinative metabolic events to continue, but restrict the seminal root protrusion [20]. Osmohardening with PEG improved seed quality of maize and leading to early germination and better performance under field conditions [133]. Polyethylene glycol (PEG) as an inert material which can prevent embryo toxicity problem during priming. The large size of PEG molecule also prevents its penetration into seed tissues, avoiding lowering the osmotic potential [134]. Seeds primed with PEG were effective in improving seed germination and seedling establishment of sorghum under unfavorable soil moisture conditions. Seed priming with PEG reduced lipid peroxidation and stabilized cell membrane, resulting in elevation of stress tolerance under drought environment [135]. Osmopriming or Osmoconditioning is the seed soaking in solutions with low water potential. In osmopriming, degree and rate of imbibitions is restricted through the exposure of seeds to low external water potential. Osmopriming can maintain the integrity of plasma membrane and gives better germination percentage [28].

Osmopriming of rice seeds enhanced starch to improve sugar availability in embryo and produced strong seedling growth. Osmopriming economically, technically and methodologically is more challenging as it yields easier and faster results than water conservation systems. Osmopriming of spinach seeds with PEG increased germination percentage, stimulated anti-oxidant defense systems and thereby induced tolerance to spinach plants. Osmopriming has advantages which include rapid and uniform germination and emergence, improved seedling growth and better stand establishment under water stress condition [64]. Osmopriming in adequate concentration of PEG improved seedling growth and germination in rice [136]. Seed osmopriming with lower dose of PEG addressed the inhibitory effects of salinity on green gram plants in terms of greater values for osmolytes accumulation, chlorophyll content and better antioxidant defense system and osmotic adjustment [137]. Priming of french bean seeds with PEG 0.1Mpa enhanced germination, emergence time and seedling vigor index [138]. Osmopriming with mannitol mitigated the inhibitory effects of salinity and drought on plant growth in chickpea. Plant produced from seeds soaked in different concentration of mannitol (2–4%) improved biomass and length of shoot and roots under saline conditions

[139]. Plants from seeds primed with mannitol had higher activities of antioxidant enzymes and minimal electrolyte leakage and malondialdehyde contents [140]. Osmoconditioning of cucumber (*Cucumis sativus* L.) seed with 0.7 M mannitol improved the rate of germination at 25°C and 15°C in water. Osmoconditioning stimulated the rate of radical extension, seedling emergence and expansion of the cotyledons and first leaf of cucumber [51].

9.3 Priming with plant growth regulators

Presoaking seeds with optimal concentrations of plant growth regulators has shown to effectively improve germination as well as growth and yield performances of various crop species crops under environmental stress conditions [141]. Growth regulators normally used for seed priming include auxins (IAA, IBA, and NAA), gibberellins (GA), kinetin, salicyclic acid, abscisic acid, ethylene and ascorbic acids. The use of plant hormones and other plant growth regulators as seed pre-sowing treatment can improve plant growth under stressful conditions [28]. Pre-soaking GA_3 in guava seeds at 100 ppm at a temperature range of 32°C/20°C resulted in a significant increase in germination to 84–88% compared to unprimed seeds [54]. Seed priming with GA_3 (100 ppm) for 24 h at low concentration and priming temperature at 15°C in cowpea increased the leaf area index, relative growth rate, crop growth rate and net assimilation rate under limited soil water conditions [142]. Rye seeds soaked with gibberellic acid increased germination percentage under water stress conditions [143].

Seed priming with ethylene minimizes the effect of high temperature on lettuce seed germination [144]. The pre-treated wheat seeds with salicylic acid improved seed germination, rate of germination and total chlorophyll content significantly under salinity levels [145]. Phytohormonal priming can augment seed germination through enhancing some enzymes such as amylase activities and protease that hydrolyzed starch and protein molecules into simple forms available for the embryo to germinate [146]. GA_3 priming has been found to improve seed germination, possibly as a result of nutrients stored inside the seeds, and to make embryos available during germination. Seed endosperm is found in the embryo through the action of other hydrolase enzymes [147]. It is very important that GA3 promotes the synthesis and production of hydrolases, especially α-amylase, which leads to seed germination. Seed priming is controlled through suppression effects of excess ABA on the expansion of embryo organs caused by inhibition of GA_3 effects on the growth of radical and hypocotyl [148]. Seeds primed with ascorbic acid improved emergence, growth and yield of maize under water deficit [149]. Seed priming with gibberellic acid induced an increase in grain yield of wheat plants, modulation of ion uptake and partitioning and hormone homeostasis under saline conditions [150].

9.4 Nutrient priming

Nutrient priming or nutripriming means soaking of seeds in nutrient solution of a specific concentration, for a certain period of time or duration prior to sowing [151]. Seed priming with nutrients (macro or micro) can increase seed nutrient content and improve seed quality for better germination, seedling establishment, plant growth, nutrient uptake and water use efficiency of several crop species. Nutrient priming is one of the methods of priming practices that includes salts like ammonium molybdate, $Mg (NO_3)_2$, $CaCl_2$, $CaSO_4$, KBr, $MgSO_4$, KH_2PO_4, $ZnSO_4$, KNO_3, sodium molybdate, KCl and NaCl in such a way that pre-emergence metabolic functions begin to prevent major outbreaks followed by seed drying at the initial humidity level [24]. In this way, the seeds are immersed in various salts that promote the germination process and subsequent emergence of seedlings even under adverse environmental conditions.

Salt priming of hot pepper seed induced salinity tolerance at seedling stage, wherein seed priming improved significantly the germination percentage and vigor index, plumule and radical length and dry weight of seedling as compared to the non-primed seeds (control) [110]. Nutrient priming in 3% KNO_3 solution for 40 h at normal room temperature increased speed of emergence, seedling vigor index, root length and shoot length over hydro priming, and control in pepper [152]. Pre-sowing seed treatment with ammonium molybdate (10^{-3} M) enhanced germination, improved vigor and growth of root system, increased drought tolerance which helped in higher nutrient uptake in cowpea crop under limited soil moisture [46]. Pepper seeds primed in 1% KNO_3 recorded the highest germination percentage as compared to non-priming [153]. Seed priming with $CaCl_2$ (2%) and sodium molybdate (100 ppm) increased the harvest index over dry seed under drought stress situation [154]. Nutrient priming is a simple and low cost agro-technique and found suitable to be recommended to the farmers owing to better synchrony of emergence and crop stand under various conditions of environment [7]. Seed ripening with $CaCl_2$ has been very successful in implanting a high salt tolerance to maize with an improved percentage of germination and biomass of plants. Plants grown from extracted seeds also raised the cellular levels of Ca^{2+}, K^+, and Na^+. Chloride content was important for maize plants raised from seeds incorporated into NaCl and KCl [155]. Effect of seed priming with KNO_3 and urea increased the seedling growth, germination percentage, germination rate and proline and protein content in maize hybrids under severe and moderate salt and drought stress [156]. Seed priming with ammonium molybdate (10^{-3} M) improved germination, stimulated growth, seed yield, biological yield and water use efficiency in cowpea under limited water supply conditions in cowpea [157]. Seed priming of chickpea seeds in a 0.05% solution of zinc sulphate ($ZnSO_4$) has been found quite effective to exhibit 19% higher seed yield and 29% more Zn concentration in seeds over that of non-primed seeds [158]. Seed priming with potassium nitrate (0.5%) recorded higher emergence, shoot length, shoot fresh weight, maximum root length and root fresh weight of dry direct-seeded rice compared with non-primed seeds [159]. Seed priming for pepper crop with osmotic solution KCl (10 mM) for 36 h improved the plant biomass, number of leaves per plant, shoot and root length, leaf area and carotenoid content under saline stress [90]. Seed priming sorghum seeds with 50 mM $Ca\ Cl_2$ enhanced the germination potential, germination rate, germination index, vigor index, root and shoot length, root and shoot fresh weight and root and shoot dry weight under salt stress condition [26].

9.5 Biopriming

Bio-priming of seeds has diverse process to stimulate morphogenesis and plant immunity, viz., production of phytohormones, induced expression of plant growth-promoting genes, mycoparasitism, increased nutrient status into the plant, antibiosis, trigger phenolic production, activation of antioxidant production, and systemic defense activation. Biopriming plays an important role in improving seed viability, germination, uniformity in emergence, plant vigor, growth and yield [160]. Biopriming agents comprehend plant growth promoting microorganisms (PGPM), Plant growth promoting bacteria (PGPB), plant growth promoting fungi (PGPF) and plant growth cyanobacteria (PGPC). PGPC are responsible for enhancing the crop growth through nitrogen fixation and release of metabolites, improving soil fertility by soil aggregation and enhancing water holding capacity [161]. Rhizoshere microbes play a very crucial role in enhanced uptake of three essential nutrients N, P and K [162]. Application of *Trichoderma* sp. through seed biopriming enhanced the enzyme activity through release of metabolites in maize plants [163]. Application of *Trichoderma harzianum* to cucumber seeds as aqueous slurry and incubated this mixture for 4 days at 20°C increased seedling emergence [164]. Slurry coating of non-primed cucumber (*C. sativus* L.) seeds with *Trichoderma harzianum* and *Trichoderma viridae* or combination of both reduced percentage of damping-off disease and increased the final emergence percentage up to 58.10% and greater seedling fresh weight [165]. Biopriming is recently used as an alternative method for controlling many seed- and soil borne pathogens [166]. Combined effect of *Pseudomonas fluorescens* and *Trichoderma harzianum* as seed biopriming resulted in significant growth of pepper seedlings [167]. Among abiotic stress amelioration by bio priming, *Trichoderma* spp. has been used in controlling salinity and drought stress in maize and wheat which exhibited better physiological and morphological parameters when compared to untreated control [168]. Biopriming with the biofungicide and clove oil 0.06% or 0.1% was an effective seed treatment to improve the vigor and relative speed of germination in hot pepper seeds [169]. Seed priming with Rhizobium + Pseudomonas at 10% for 12 h recorded significantly higher germination percent and speed of germination, and seedling vigor in chickpea [170]. Application of *Pseudomonas aureofaciens* through drum priming system enhanced the stand establishment in tomato [171]. The results showed that seed inoculation with plant growth promoting rhizobacteria had significantly effects on grain yield, grain 1000 weight, number of grains per plant, plant height and all of grain filling parameters such as grain filling period, rate and effective grain filling period inlentil [172]. The technique of biopriming to document using two strains including *Azospirillum brasilense* and *Bacillus amyloliquefaciens* increased drought tolerance in wheat plants through regulation of genes related to stress. Biopriming, an amalgamation of seed priming with application of plant beneficial fungi and bacteria, can remarkably improve seed germination and emergence, seedling establishment, crop growth, and yield parameters under normal and stress conditions [173]. *Thallasso bacillus denorans* and *Oceano bacillus kapialis* isolates from salt mine showing halophillic behavior enhanced the growth of fine rice variety under varying salinity concentrations and exhibited improvement in morphological and physiological parameters after 15 and 28 days, respectively, when applied through biopriming [174]. Bioprimping of *Medicago truncatula* seeds with *Bacillus* spp. Improved seed germination and seedling biomass and at the molecular level reflected in the up regulation of genes involved in DNA damage repair and antioxidant defense [88].

9.6 Solid matrix priming

In solid matrix priming (SMP) or matrix conditioning, solid or semi solid medium is used as a substitute in place of liquid medium. This technique is accomplished by mixing seeds with a solid or semi solid medium and specified amount of water. In solid matrix priming, a small quantity of seed and solid particles are used. During solid matrix priming, water is slowly delivered to seeds and thus, slow or controlled imbibition occurs, allowing cell repair mechanisms to function [28]. Predominant solid matrices are exfoliated vermiculate, expanded calcined clay, bituminous coal, sodium polypropionate gel or synthetic calcium silicate. Solid matrix priming using saw dust, ground charcoal, green gram seeds responded favorably to shorten incubation periods. The longer incubation periods and higher water levels were harmful to the seeds because they encouraged fungal growth [175].

10. Conclusion and future perspective

Seed priming emerges as a reassuring technology for combating abiotic stress in crops and alleviating the detrimental effects of abiotic stress without much influencing its fitness. Seed priming technique is innovative, cheap and simple to apply at farmer's field conditions. Oxidative stress, temperature extremes, salinity, and drought are associated and frequently induce similar type of damage. Seed priming stimulates signaling pathways earlier and enhances plant defense responses. Experimental results reveal that improved germination and vigorous growth of seedlings occur in early seed by combining stored nutrients and using genes responsible for the synthesis of essential enzymes. Priming is also capable of repairing damage that occurs inside the seed. Seed priming effects on early stage of germination, and it modulates the DNA replication, transcription, and translation. Storage and short shelf life of the primed seeds are a limitation of this technology. There is a need to standardize suitable priming methods in different crops to combat abiotic stress sustainably. Seed priming may indeed be considered as a valuable strategy to improve stand establishment under detrimental agro-climatic conditions (rainfed, dry farming and dry land farming regions) with enhanced yield, increased tolerance to stress situations, enhanced crop competitiveness against weeds, increased resistance against diseases and increased water use efficiency. In an outline, seed priming acts as an important criterion for the induction of tolerance in plants against a wide range of abiotic stresses. However, more investigation will be needed in unraveling the mechanism of plant growth regulators and their substitutes, especially with stress-responsive genes.

Author details

Melekote Nagabhushan Arun[1*], Shibara Shankara Hebbar[2], Bhanuprakash[2], Thulasiram Senthivel[3], Anil Kumar Nair[2], Guntupalli Padmavathi[4], Pratima Pandey[5] and Aarti Singh[1]

1 Department of Agronomy, Indian Council of Agricultural Research-Indian Institute of Rice Research (ICAR–IIRR), Hyderabad, India

2 Department of Vegetable Crop, ICAR-Indian Institute of Horticultural Research, Bangalore, India

3 Department of Agriculture and Animal Husbandry, Gandhigram Rural Institute, Dindigul, India

4 Department of Plant Breeding, ICAR-Indian Institute of Rice Research, Hyderabad, India

5 Department of Seed Science, Nepal Agricultural Research Council, Kathmandu, Nepal

*Address all correspondence to: arun_tulasi2011@yahoo.in

References

[1] Kooyers NJ. The evolution of drought escape and avoidance in natural herbaceous populations. Plant Science. 2015;**234**:155-162

[2] Zhao TJ, Liu Y, Yan YB, Feng F, Liu WQ, Zhou HM. Identification of the amino acids crucial for the activities of drought responsive element binding factors (DREBs) of *Brassica napus*. FEBS Letters. 2007;**581**:3044-3050

[3] Bewley JD, Black M. Seeds: Physiology of Development and Germination. 2nd ed. New York: Plenum Press; 1994. DOI: 10.1007/978-1-4899-1002-8

[4] Zhang Y, Pan J, Huang X, Gua D, Lou H, Hou Z, et al. Differential effects of a post-anthersis heat stress on wheat (*Triticum aestivum* L.) grain proteomic determine iTRAQ. Science Reporter. 2017;7:3468

[5] Hamidi H, Safarnejad A. Effect of drought stress on alfalfa cultivars (*Medicago sativa* L.) in germination stage. American-Eurasian Journal of Agricultural and Environmental Sciences. 2010;**8**(6):705-709

[6] Jisha KC, Puthur JT. Halopriming of seeds imparts tolerance to NaCl and PEG induced stress in *Vigna radiata* (L.) Wilczek varieties. Physiology and Molecular Biology of Plants. 2014;**20**(3): 303-312

[7] Sedghi M, Nemati A, Amanpour-Balaneji B, Gholipouri A. Influence of different priming materials on germination and seedling establishment of milk thistle (*Silybum marianum*) under salinity stress. World Applied Sciences Journal. 2010;**11**(5):604-609

[8] Ahmad I, Khaliq T, Ahmad A, Basra SMA, Hasnain Z, Ali A. Effect of seed priming with ascorbic acid, salicylic acid and hydrogen peroxide on emergence, vigor and antioxidant activities of maize. African Journal of Biotechnology. 2012;**11**:1127-1137

[9] Liu Y, Li P, Xu GC, Xiao L, Li ZB. Growth, morphological and physiological responses to drought stress in *Bothrochloa ischaemum*. Frontiers in Plant Science. 2017;**8**:230

[10] Alzahrani FO. Metabolic of osmo protectants to elucidate the mechanism of salt stress tolerance in crop plants. Planta. 2021;**253**:1-17

[11] Blum A. Osmotic adjustment is a prime drought stress adaptive engine in support of plant production: Osmotic adjustment and plant production. Plant Cell Environment. 2017;**40**:4-10

[12] Harris D, Joshi A, Khan PA, Gothkar P, Sodhi PS. On-farm seed priming in semi-arid agriculture development and evaluation in maize, rice and chickpea in India using participatory methods. Experimental Agriculture. 1999;**35**:15-29

[13] Ashraf M, Afaf YR, Qureshi MS, Sarwar G, Naqvi MH. Salinity induced changes in amylase and protease activities and associated metabolism in cotton varieties during germination and early seedling growth stages. Acta Physiologiae Plantarum. 2002;**24**:37-44

[14] Khajeh-Hosseini M, Powell AA, Bimgham IJ. The interaction between salinity stress and seed vigour during germination of soybean seeds. Seed Science and Technology. 2003;**31**:715-725

[15] Hussian S, Zhang J, Zhong C, Zhu L, Cao X, Yu S. Effect of salt stress on rice growth, development characteristics and the regulating ways: A review. Journal of Integrative Agriculture. 2017;**16**(11):2357-2374

[16] Vollenweider P, Gunthardt-Goerg MS. Diagnosis of abiotic and biotic stress

factors using the visible symptoms in foliage. Environmental Pollution. 2005;**137**:455-465

[17] Wahid A, Gelani S, Ashraf M, Foolad MR. Heat tolerance in plants: An overview. Environmental and Experimental Botany. 2007;**61**(3): 199-223

[18] Zinn KE, Tunc-Ozdemir M, Harper JF. Temperature stress and plant sexual reproduction: Uncovering the weakest links. Journal of Experimental Botany. 2010;**61**:1959-1968

[19] Hasanuzzaman M, Nahar K, Fujita M, Ahmad P, Chandna R, Prasad M, et al. Enhancing plant productivity under salt stress: Relevance of poly-omics. In: Salt Stress in Plants. New York, NY: Springer; 2013. pp. 113-156. DOI: 10.1007/978-1-4614-6108-1_6

[20] Ashraf M, Foolad MR. Pre-sowing seed treatment—A shotgun approach to improve germination, plant growth, and crop yield under saline and non-saline conditions. Advance of Agronomy. 2005;**88**:223-276

[21] Nleya T, Ball RA, Vandenberg A. Germination of common bean under constant and alternating cool temperatures. Canadian Journal of Plant Sciences. 2005;**85**(3):577-585

[22] Sivritepe HO. The effects of osmotic conditioning treatments on salt tolerance of Onion seeds. In: 3rd National Symposium on Vegetable Production; Isparta, Turkey. 2000. pp. 475-481

[23] Bradford KJ. Manipulation of seed water relations via osmotic priming to improve germination under stress conditions. HortScience. 1986;**21**: 1105-1112

[24] McDonald MB. In: Bewley MJD, editor. Seed Priming, Black, Seed Technology and Its Biological Basis. Sheffield Academic Press, Sheffield, UK; 2000. pp. 287-325

[25] Savvides A, Ali S, Tester M, Fotopoulas V. Chemical priming of plants against multiple abiotic stresses: Mission possible? Trends in Plant Science. 2016;**21**:329-340

[26] Chen X, Zhang R, Xing Y, Jiang B, Li B, Xu X. The efficacy of different seed priming agents for promoting sorghum germination under salt stress. PLoS One. 2021;**16**(1):e0245505. DOI: 10.1371/journal.pone.0245505

[27] Fujita M, Fujita Y, Noutoshi Y, Takahashi F, Narusaka Y, Yamaguchi-Shinozaki K, et al. Crosstalk between abiotic and biotic stress responses: A current view from the points of convergence in the stress signaling networks. Current Opinion Plant Biology. 2006;**9**:436-442

[28] Jisha KC, Vijayakumari K, Puthur JT. Seed priming for abiotic stress tolerance: An overview. Acta Physiologiae Plantarum. 2013;**35**:1381-1396

[29] Siri B, Vichitphan K, Kaewnaree P, Vichitphan S, Klanrit P. Improvement of quality, membrane integrity and antioxidant systems in sweet pepper (*Capsicum annuum* Linn) seeds affected by osmopriming. Australian Journal of Crop Science. 2013;**7**(13):2068-2073

[30] Harris D, Raghuwanshi BS, Gangwar JS, Singh SC, Joshi KD, Rashid A, et al. Participatory evaluation by farmers of on-farm seed priming in wheat in India. Nepal and Pakistan. Experimental Agriculture. 2001;**37**: 403-415

[31] Khan FA, Maqbool R, Narayan S. Reversal of age-induced seed deterioration through priming in vegetable crops—A review. Advance in Plants and Agriculture Research. 2016;**4**(6):403-411

[32] Sano N, Seo M. Cell cycle inhibitors improve seed storability after priming treatments. Journal of Plant Research. 2019;**132**:263-271

[33] Arun MN, Bhanuprakash K, Shankara Hebbar S, Senthivel T. Effect of seed priming with various chemicals on germination and vigour in cowpea (*Vigna unguiculata* (L.) Walp). Progressive Research—An International Journal. 2016;**11**:2113-2117

[34] Salisbury FB, Ross CW. Plant Physiology. Belmont, CA: Wadsworth; 1992. pp. 357-407, 531-548

[35] Hussain A, Rizwan M, Ali Q, Ali S. Seed priming with silicon nano particles improved the biomass and yield while reduced the oxidative stress and cadmium concentration in wheat grains. Environmental Science and Pollution Research. 2019;**26**(8):7579-7588

[36] Heydecker W, Coolbear P. Seed treatments for improved performance; survey and attempted prognosis. Seed Science and Technology. 1977;**5**:353-425

[37] Cantliffe DJ. Priming of lettuce seed for early and uniform emergence under conditions of environmental stress. Acta Horticulturae. 1981;**122**:29-38

[38] Ghassemi-Golezani K, Jabbarpour S, Zehtab-Salmasi S, Mohammadi A. Response of winter rapeseed (*Brassica napus* L.) cultivars to salt priming of seeds. African Journal of Agricultural Research. 2010;**5**:1089-1094

[39] Rajpar I, Kanif YM, Memon AA. Effect of seed priming on growth and yield ofwheat (*Triticum aestivum* L.) under non-saline conditions. International Journal of Agricultural Research. 2006;**1**:259-264

[40] Anju BR, Sheeja R. Seed priming: An approach towards agricultural sustainability. Journal of Applied and Natural science. 2019;**11**(1):227-234

[41] Anese S, Da Silva EAA, Davide AC, Rocha Faria JM, Soares GCM, Matos ACB, et al. Seed priming improves endosperm weakening, germination, and subsequent seedling development of *Solanum lycopersicum* St. Hil. Seed Science and Technology. 2011;**39**:125-139

[42] Sarika G, Basavaraju GV, Bhanuprakash K, Chaanakeshava V, Paramesh R, Radha BN. Investigation on seed viability and vigour of aged seed by priming in french bean. Vegetable Science. 2013;**40**:169-173

[43] Yari L, Aghaalikhani M, Khazaei F. Effect of seed priming duration and temperature on seed germination behavior of bread wheat (*Triticum aestivum* L.). ARPN Journal of Agricultural and Biological Science. 2010;**5**(1):5-8

[44] Kausar M, Mahmood T, Basra SMA, Arshad M. Invigoration of low vigor sunflower hybrids by seed priming. International Journal of Agriculture and Biology. 2009;**11**:521-528

[45] Assefa MK, Hunje R. Standardization of seed priming duration in soybean [*Glycine max* (L.) Merill]. Seed Research. 2011;**39**:1-4

[46] Arun MN, ShankaraHebbar S, Bhanuprakash K, Senthivel T. Effect of seed priming on phenology, growth and yield of vegetable cowpea (*Vigna unguiculata* (L.) Walp) under different intensities of moisture stress. Progressive Research—An International Journal. 2016;**11**:2118-2122

[47] Bhanuprakash K, Yogeesha HS, Arun MN, Naik LB. Effect of priming and ageing on seed germination and vigour in papaya (*Carica papaya*). Indian Journal of Agricultural Sciences. 2009;**79**(4):295-297

[48] Parashar A, Verma SK. Effect of pre-sowing seed soaking in gibberellic

acid, duration of soaking temperatures and their interaction on seed germination and early seedling growth of wheat under saline conditions. Plant Physiology and Biochemistry. 1988; **15**(2):189-197

[49] Chen K. Antioxidants and dehydrin metabolism associated with osmopriming-enhanced stress tolerance of germinating spinach (*Spinacia oleracea* L. cv. Bloomsdale) seeds. In: Graduate Theses and Dissertations, Paper 10471. USA: Iowa State University; 2011. DOI:10.31274/ETD-180810-355

[50] Barthwal P, Prabha D. Effect of ageing and hormonal priming on different physiological attributes on French beans (*Phaseolus vulgaris*). Advances in Agriculture and Natural Sciences for Sustainable Agriculture. 2018;**5**:76-79

[51] Amooaghaie R, Vaviland M. The combined effect of gibberallic acid andlong time osmoprimng on seed germination and subsequent seedling growth of *Klussia odoratissima* Mozaff. African Journal of Biotechnology. 2011;**10**(66):14873-14880

[52] Renugadevi J, Natarajan N, Srimathi P. Efficacy of botanicals in improving the seeds and seedling quality characteristics of cluster bean. Legume Research—An International Journal. 2008;**31**:164-168

[53] Khan FA, Maqbool R, Narayan S, Bhat SA, Narayan R, Khan FU. Reversal of age-induced seed deterioration through priming in vegetable crops—A review. Advances in Plants and Agriculture Research. 2016;**4**(6):403-410

[54] Bhanuprakash K, Yogeesha HS, Vasugi C, Arun MN, Naik LB. Effect of pre-soaking treatments and temperature on seed germination of guava (*Psidium guajava* L.). Seed Science and Technology. 2008;**36**:792-794

[55] Bewley JD. Seed germination and dormancy. The Plant Cell. 1997;**9**(7): 1055-1066

[56] Bray CM. Biochemical processes during the osmopriming of seeds. In: Kigel J, Galili G, editors. Seed Development and Germination. Marcel Dekker: New York; 1995. pp. 767-789

[57] Ibrahim EA. Seed priming to alleviate salinity stress in germinating seeds. Journal of Plant Physiology. 2016;**192**:38-46

[58] Chen K, Arora R. Dynamics of the antioxidant system during seed osmopriming, post priming germination, and seedling establishment in spinach (*Spinacia oleracea*). Plant Science. 2011;**180**:212-220

[59] Paparella S, Araújo SS, Rossi G, Wijayasinghe M, Carbonera D, Balestrazzi A. Seed priming: State of the art and new prospective. Plant Cell Reports. 2015;**34**:1281-1293

[60] Bradford KJ, Bewley JD. Seeds: Biology, technology and role in agriculture. In: Chrispeels MJ, Sadava DE, editors. Plants, Genes and Crop Biotechnology. 2nd ed. Boston: Jones and Bartlett; 2002. pp. 210-239

[61] Bose B, Tandon A. Effect of magnesium nitrate on metabolism in germinating maize seeds. Indian Journal of Plant Physiology. 1991;**34**:69-71

[62] Pandita VK, Anand A, Nagarajan S. Enhancement of seed germination in hot pepper following presowing treatments. Seed Science and Technology. 2007;**35**:282-290

[63] Catusse J, Meinhard J, Job C, Strub JM, Fischer U, Pestova E, et al. Proteomics reveals potential biomarkers of seed vigor in sugar beet. Proteomics. 2011;**11**:1569-1580

[64] Chen Y, Cui J, Li G, Yuan M, Zhang Z, Yuan S, et al. Effect of salicylic

acid on the antioxidant system and photosystem II in wheat seedlings. Biology of Plant. 2016;**60**:139-147

[65] Li X, Cai J, Liu F, Dai T, Cao W, Jiang D. Cold priming drives the sub-cellular antioxidant systems to protect photosynthetic electron transport against subsequent low temperature stress in winter wheat. Plant Physiology and Biochemistry. 2014;**82**:34-43

[66] McDonald MB. Seed deterioration: Physiology, repair and assessment. Seed Science and Technology. 1999;**27**:177-237

[67] Sen A, Jos TP. Influence of different seed priming technique on oxidative and antioxidative responses during the germination of *Oryza sativa* varieties. Physiology and Molecular Biology of Plants. 2020;**26**(3):551-556

[68] Varier A, Vari AK, Dadlani M. The sub cellular basis of seed priming. Current Science. 2010;**99**(4):450-456

[69] Farooq M, Basra SMA, Hussain M, Rehman H, Saleem BA. Incorporation of polyamines in the priming media enhances the germination and early seedling growth in hybrid sunflower (*Helianthus annuus* L). International Journal of Agriculture and Biology. 2007;**9**:868-872

[70] Sivritepe HO, Dourado AM. The effect of priming treatments on the viability and accumulation of chromosomal damage in aged pea seeds. Annals of Botany. 1995;**75**:165-171

[71] Dahal P, Bradford KJ, Jones RA. Effects of priming and endosperm integrity on seed germination rates of tomato genotypes. II. Germination at reduced water potential. Journal of Experimental Botany. 1990;**41**: 1441-1453

[72] Dollypan BRN. Mid-storage and pre-sowing seed treatments for lettuce and carrot. Scientia Horticulturae. 1985;**33**:1026-1027

[73] Villiers TA, Edgcumbe DJ. On the cause of seed determination in dry storage. Seed Science and Technology. 1975;**3**:761-774

[74] Panda D, Mondal S. Seed enhancement for sustainable agriculture: An overview of recent trends. Plant Archives. 2020;**20**(1):2320-2332

[75] Ali M, Hayat M, Ahmad H, Ghani MI, Amin B, Atif MJ, et al. Priming of *Solanum melongena* L. seeds enhancing germinatiuon, alters antioxidant enzymes, modulated ROS and improves early seedling growth: Indicating aqueous garlic extract as seed-priming bio-stimulant for egg plant production. Applied Sciences. 2019;**9**:1-18

[76] Sananda M, Bose B. Kinetics studies on α-amylase extracted from germinating wheat endosperm of primed and non-primed seeds. Indian Journal of Agricultural Biochemistry. 2012;**25**:137-141

[77] Lee SS, Kim JH, Hong SB, Yun SH, Park EH. Priming effect of rice seeds on seedling establishment under adverse soil conditions. Korean Journal of Crop Science. 1998;**43**:194-198

[78] Chen K, Arora R. Priming memory invokes seed stress-tolerance. Environmental and Experimental Botany. 2013;**94**:33-45

[79] Jyoti MCP. Seed deterioration: A review. International journal of life science botany and pharmaceutical. Research. 2013;(3):374-385

[80] Manonmani V, Ameer Junaithal Begum M, Jayanthi M. Halo priming of seeds. Research Journal of Seed Science. 2014;**7**:1-13

[81] Georghiou K, Thanos CA, Passam HC. Osmoconditioning as a means of counteracting the ageing of pepper seeds during high-temperature storage. Annals of Botany. 1987;**60**:279-285

[82] Panayotov N, Aladjadjiyan A. Effect of long-term storage of pepper (*Capsicum annuum* L.) seeds on their viability measured by selected thermodynamic parameters. Acta Scientiarum Polonorum Hortorum Cultus. 2014;**13**(2):151-162

[83] Arun MN, Shankara Hebbar S, Bhanuprakash K, Senthivel T, Nair AK, Pratima PD. Biochemical investigation on vigour enhancement in fresh and aged seeds upon seed priming in cowpea (*Vigna unguiculata* (L.) Walp). Legume Research—An International Journal. 2021;**44**(12):1497-1505. DOI: 10.18805/LR-4476

[84] Srinivasan K, Jain SK, Saxena S, Radhamani J, Uprety M. Seed priming and fortifcation. Seed Research. 2009;**37**(1&2):1-13

[85] Ermiş S, Kara F, Özden E, Demir I. Solid matrix priming of cabbage seed lots: Repair of ageing and increasing seed quality. Journal of Agricultural Sciences. 2016;**22**:588-595

[86] Thornton JM, Collins ARS, Powell AA. The effect of aerated hydration on DNA synthesis in embryos of *Brassica oleracea* L. Seed Science Research. 1993;**3**:195-199

[87] Arun MN, Bhanuprakash K, Shankara Hebbar S, Senthivel T. Effect of seed priming on biochemical parameters and seed germination in cowpea (*Vigna unguiculata* (L.). Walp). Legume Research—An International Journal. 2017;**40**(3):562-570

[88] Chiari F, Ajay S, Anjali S, Alma S, Prasad V, Anca M. Hydropriming and biopriming improve *Medicago truncatula* seed germination and upgrade DNA repair and antioxidant genes. Genes. 2020;**11**:1-15

[89] Lutts S, Benincasa P, Wojtyla L, Kubala S, Pace R, Lechowska K, et al. Seed priming: New comprehensive approaches for an old empirica technique, new challenges in seed biology. In: Susana Araújo S, Balestrazzi A, editors. Basic and Translational Research Driving Seed Technology. Open, Rijeka: In Tech; 2016. pp. 1-47

[90] Aloui H, Eymen EM, Cherif H. Seed priming to improve seedling growth of pepper cultivars exposed to salt concentrations. International Journal of Vegetable Science. 2017;**23**:489-507

[91] Gallardo K, Job C, Groot SPC, Puype M, Demol H, Vandekerekhove J, et al. Proteomics of Arabidopsis seed germination and priming. In: Nicholas G, editor. The Biology of Seeds: Recent Advances. CABI: Cambridge; 2002. pp. 199-209

[92] Kibinza S, Bazin J, Bailly C, Farrant JM, Corbineau F, El-Maarouf-Bouteau H. Catalase is a key enzyme in seed recovery from ageing during priming. Plant Science. 2011;**181**:309-315

[93] Wojtyla Ł, Lechowska K, Kubala S, Garnczarska M. Different modes of hydrogen peroxide action during seed germination. Frontiers in Plant Science. 2016;7:66. DOI: 10.3389/fpls.2016.00066

[94] Bailly C. Active oxygen species and antioxidants in seed biology. Seed Science Research. 2004;**14**:93-107

[95] Coolbear P, Slater RJ, Bryant JA. Changes in nucleic acid levels associated with improved germination performance of tomato seeds after low-temperature pre-sowing treatments. Annals of Botany. 1990;**65**:187-195

[96] Kaya G, Demir I, Tekin A, Yasar F, Demir K. The effect of priming application on germination, fatty acids, sugar content and enzyme activity at the temperature of stress of pepper seeds. Journal of Agricultural Science. 2010;**16**:9-16

[97] Job D, Capron I, Job C, Dacher F, Corbineau F, Coˆme D. Identification of germination-specific protein markers and their use in seed priming technology. In: Black M, Bradford KJ, Va'zquez-Ramos J, editors. Seed Biology: Advances and Applications. Wallingford, UK: CAB International; 2000. pp. 449-459

[98] Cheng J, Wang L, Zeng P, He Y, Zhou R, Zhang H, et al. Identification of genes involved in rice seed priming in the early imbibition stage. Plant Biology. 2017;**19**:61-69

[99] Kubala S, Garnczarska M, Wojtyla L, Clippe A, Kosmala A, Zmienko A, et al. Deciphering priming-induced improvement of rapeseed (*Brassica napus* L.) germination through an integrated transcriptomic and proteomic approach. Plant Science. 2015;**231**:94-113

[100] Iqbal M, Ashraf M. Seed treatment with auxins modulates growth and ion partitioning in salt-stressed wheat plants. Journal of Integrative Plant Biology. 2007;**49**:1003-1015

[101] Nakao Y, Asea G, Yoshino M, Kojima N, Hanada H, Miyamoto K, et al. Development of hydropriming techniques for sowing seeds of upland rice in Uganda. American Journal of Plant Sciences. 2018;**9**:2170-2182

[102] Bhusal D, Thakur DP. Seed hydropriming technique in cereal crops: A review. Reviews in Food and Agriculture. 2020;**1**(2):85-88

[103] Elisabeth Zida P, James Neya B, RomainSoalla W, Sereme P. Ole Sogaard Lund. Hydropriming of pearl millet in northern and central Burkino Faso applying six hours of soaking and overnight drying of seeds. African Journal of Agricultural Research. 2017;**12**(149):3441-3446

[104] Sharma R, OmvatiVerma S. Response to pre-sowing seed treatment on germination indices, seeding growth and enzymatic activities of chickpea (*Cicer arietinum* L.) seed. International Journal of Ecology and Environmental Sciences. 2021;**3**(1):405-410

[105] Catiempo RL, Photchanachai S, Bayogan ERV, Chalermchai W. Impact of hydropriming on germination and seedling establishment of sunflower seeds at elevated temperature. Plant Soil and Environment. 2021;**67**:491-498

[106] Elien Lemmens, Lomme JD, Neils De Brier, Wannes L DeMan, Marice De Proft, Els Primsen, Jan AD. The impact of hydropriming and Osmopriming on seedling characteristics, plant hormone concentrations, activity of selected hydrolytic enzymes and cell wall and phytate hydrolysis in sprouted wheat. ACS Omega 2019;4:22089-22100. Available from: http//pubsacs.org/journal/acsdf

[107] Pirosteh-Anosheh H, Hashemi SE. Priming, a promising practical approach to improve seed germination and plant growth in saline soils. Asian Journal of Agriculture and Food Sciences. 2020;**8**(1):6-10

[108] Sumar Varshini P, Bayyapu Reddy K, Radhika K, SaidaNaik V. Effect of concentration and duration of osmopriming on germination and vigour of aged seed of chickpea. International Journal of Current Microbiology and Applied Sciences. 2018;**7**(10):2410-2421

[109] Cokkizgin A, Girgel U, Cokkizgin H. Mannitol effects on germination of broad bean (*Vicia faba* L.) seeds. Forestry Research and Engineering International Journal. 2019;**3**(1):20-22

[110] PiyadaTheerakulpisut NK, BunikaPanwang S. Priming alleviated salt stress effects on rice seedlings by improving Na^{+}/K^{+} and maintaining membrane integrity. International Journal of Plant Biology. 2016;7:53-58

[111] Shelar A, Singh AV, Maharjan RS, Laux P, Luch A, Gemmati D, et al. Sustainable agriculture through multidisciplinary seed nanopriming: Prospects of opportunities and challenges. Cell. 2021;**10**:6-22. DOI: 10.3390/cells10092428

[112] Dhillon BS, Kumar V, Sagwal P, Kaur N, Singh Mangat G, Singh S. Seed priming with potassium nitrate and gibberellic acid enhances the performance of dry direct seeded Rice (*Oryza sativa* L.) in North-Western India. Agronomy. 2021;**11**:849. DOI: 10.3390/agronomy11050849

[113] Mangena P. Effect of hormonal seed priming on germination, growth, yield and biomass allocation in soybean grown under induced drought stress. Indian Journal of Agricultural Research. 2020;**54**:441. DOI: 10.18805/IJARe.A-441

[114] Zhao T, Ding X, Xiao Q, Han Y, Zhu S, Chen J. IAA improves the germination and seedling growth in cotton via regularing the endogenous phytohormones and enhancing the sucrose metabolism. Industrial Crops and Products. 2020;**1555**:1-9. DOI: 10.1016/j.indcrop.2020.112788

[115] Khanmdi N, Nabipour M, Roshanfekr H, Rahmana A. Effect of seed priming and application and application of cytokinin and auxin on growth and grain yield of wheat under Alvay climatic condition. Iranian Journal of Crop Science. 2019;**21**(1):31-44

[116] Ahmad F, Iqbal S, Khan MR, Abbas MW, Ahmad J, Nawaz H, et al. Influence of seed priming with salicylic acid on germination and early growth of sesame. Pure and Applied Biology. 2019;**8**(2):1206-1213

[117] Prasad SR, Kamble UR, Sripathy KV, Udaya Bhaskar K, Singh DP. Seed bio-primingfor biotic and abiotic stress management. In: Singh DP, Singh HB, Prabha R, editors. Microbial Inoculants in Sustainable Agricultural Productivity: Vol. 1: Research Perspectives. New Delhi: Springer; 2016. pp. 211-228

[118] Sathiyanarayanan G, Suvarna G, Baradhan G, Prakash G. Effect of various seed halo-priming treatments on seed yield and quality in maize. Plant Archives. 2019;**19**(1):377-383

[119] Singhal RK, Pandey S, Bose B. Seed priming with $Mg(NO_3)_2$ and $ZnSO_4$ salts triggers physio-biochemical and antioxidant defense to induce water stress adaptation in wheat (*Triticum aestivum* L.). Plant Stress. 2021;**2**:1-12. DOI: 10.1016/j.stress2021.10003

[120] Majda C, Khalid D, Aziz A, Rachid B, Badr A-S, Lotfi A, et al. Nutrient-priming as an efficient means to improve the agronomic performance of molybdenum in common bean (*Phaseolus vulgaris* L.). Science of the Total Environment. 2019;**661**:654-663

[121] Krishnaprabu S. Response of green gram to pre-sowing seed priming chemicals. International Journal of Pure and Applied Biosciences. 2018;**6**(6): 455-458

[122] Bisht N, Mishra SK, Chauhan PS. *Bacillus amyloliquifaciens* inoculation alters physiology of rice (*Oryza sativa* L. var.IR36) through modulating carbohydrate metabolism to mitigate stress induced by nutrient starvation. International Journal of Biological Macromolecules. 2020;**15**(143):935-951

[123] Li H, Yue H, Li L. Seed biostimulant *Bacillus* sp. MGW8 improves the salt tolerance of maize during seed germination. AMB Express. 2021;**11**:74. DOI: 10.1186/s13568-021-012371

[124] Mahmood A, Kataoka R. Potential of biopriming in enhancing crop productivity and stress tolerance. In: Rakshit A, Singh H, editors. Advances in

Seed Priming Singapore. 2018. pp. 127-145. DOI: 10.1007/978-981-13-0032-5_9

[125] Sumbul A, Ansari RA, Rizvi R, Mahmood I. A potential bio-fertilizer for soil and plant health management. Saudi Journal of Biological Sciences. 2020;**27**(12):3634-3640

[126] Fukami J, Cerezini P, Hungria M. Azospirillium: Benefits that go far beyond biological nitrogen fixation. AMB Express. 2018;**8**:73. DOI: 10.10.11186/s13568-018-0608

[127] Ahammad KU, Rahman MM, Ali MR. Effect of hydro priming method on maize (*Zea mays*) seedling emergence. Bangladesh Journal of Agricultural Research. 2014;**39**(1):43-150

[128] Patra SS, Mehera B, Rout S, Tomar SS, Singh M, Kumar R. Effect of hydro-priming and different sowing dates on growth and yield attributes of wheat (*Triticum aestivum* L.). Journal of Applied and Natural Science. 2016;**8**(2):971-980

[129] Nakao Y, Asea G, Minoru Y, Nobuki K, Hiroyuki H, Kisho M, et al. Development of hydro priming techniques for sowing seeds of upland rice in Uganda. American Journal of Plant Sciences. 2018;**9**(11):2170-2182

[130] Langeroodi ARS, Noora R. Seed priming improves the germination and field performance of soybean under drought stress. The Journal of Animal and Plant Sciences. 2017;**27**(5):1611-1621

[131] Banjobpudsa S, Sripichitt A, Sarutayophat T. The effect of pre-sowing treatments on germination and vigour of upland rice (*Oryza sativa* L.). International Journal of Agricultural Technology. 2017;**13**:1343-1353

[132] Heydecker W, Higgins J, Gulliver RL. Accelerated germination by osmotic seed treatment. Nature. 1973;**5427**:42-44

[133] Jalal R, Khan AZ, Anwar S, Ahmad J, Safia B, Ahmad F, et al. Influence of different pre-sowing invigoration techniques on early growth of different maize hybrids. International Journal of Biosciences;**15**(2):370-379

[134] Brocklehurst PA, Dearman J. Interaction between seed priming treatments and nine seed lots of carrot, celery and onion. II. Seedling emergence and plant growth. Annals of Applied Biology. 2008;**102**:583-593

[135] Zhang F, Yu J, Johnston CR, Wang Y, Zhu K, Lu F, et al. Seed priming with polyethylene glycol induces physiological changes in sorghum(*Sorghum bicolour* L. Moench) seedlings under suboptimal soil moisture environments. PLoS One. 2015;**10**(10):e0140620. DOI: 10.1371/journal.pone.0140620

[136] Sun L, Zhou Y, Wang C, Xiao M, Tao Y, Xu W. Screening and identification of sorghum cultivars for salinity tolerance during germination. Scientia Agricultura Sinica. 2012;**45**(9):1714-1722

[137] Saha P, Chatterjee P, Biswas AK. NaCl pretreatment alleviates salt stress by enhancement of antioxidant defense system and osmolyte accumulation in mungbean (*Vigna radiate* (L.) Wilczek). Indian Journal of Experimental Biology. 2010;**48**:593-600

[138] Bhanuprakash K, Yogeesha HS, Arun MN, Naik LB. Studies on seed quality in relation to ageing and priming in french beans cv.Arka komal. In: Proceeding of the Fourth International Food Legume Research Conference—IV New Delhi India. 2005. pp. 771-776

[139] Kaur D, Grewal S, Kaur J, Singh S. Differential proline metabolism in vegetative and reproductive tissues determine drought tolerance in chickpea. Biology of Plant. 2017;**61**(2):359-366

[140] Matias JR, Ribeiro RC, Arago CA, Araujo GGL, Dantas BF. Physiological

changes in osmo and hydro-primed cucumber seeds germinated in biosaline water. Journal of Seed Science. 2015;**37**(1):7-15

[141] Bahrani A, Pourreza J. Gibberellin acid and salicylic acid effects on seed germination and seedlings growth of wheat under salt stress condition. World Applied Science Journal. 2012;**18**(5): 633-641

[142] Arun MN, Shankara Hebbar S, Bhanuprakash K, Senthivel T, Nair AK, Pratima PD. Influence of seed priming and different irrigation levels on growth parameters of cowpea (*Vigna unguiculata* (L.) Walp). Legume Research—An International Journal. 2018;**43**(1):99-104

[143] Ansari O, Azadi MS, Sharif-Zadeh F, Younesi E. Effect of hormone priming on germination characteristics and enzyme activity of mountain rye (*Secale montanum*) seeds under drought stress conditions. Journal of Stress Physiology and Biochemistry. 2013;**9**:61-71

[144] Nascimento WM, Cantliffe DJ, Huber DJ. Ethylene evolution and endo-beta-mannanase activity during lettuce seed germination at high temperature. Scientia Agricola. 2004;**61**(2):156-163

[145] Azeem M, Qasim M, Abbas MW, Tayyab SB, Yousuf M, Ali H. Salicyclic acid seed priming modulates some biochemical parameters to improve germination and seedling growth of salt stressed wheat (*Triticum aestivum* L.). Pakistan Journal of Botany. 2019;**51**(2):1-7

[146] Miransari M, Smith DL. Plant hormones and seed germination. Environmental and Experimental Botany. 2014;**99**:110-121

[147] Abu-Muriefah SS. Phytohormonal priming improves germination and antioxidant enzymes of soybean (*glycine max* L.) seeds under lead (Pb) stress. Bioscience Research. 2017;**14**(1):42-56

[148] Voegele A, Linkies A, Muller K, Leubner-Metzger G. Members of the GIBBERELLIN receptor gene family GID1 (GIBBERELLIN INSENSITIVEDWARF1) play distinct roles during *Lepidium sativum* and *Arabidopsis thaliana* seed germination. Journal of Experimental Botany. 2011;**155**:1851-1870

[149] Farooq M, Hussain M, Wakeel A, Kadambot HMS. Salt stress in maize: Effects, resistance mechanisms and management—A review. Agronomy for Sustainable Development. 2015;**35**:461-448

[150] Iqbal M, Raja NI, Yasmeen F, Hussain M, Ejaj M, Shah MA. Impacts of heat stress on wheat: A critical review. Advances in Crop Sciences and Technology. 2017;**5**(1):01-09

[151] Shivay YS, Singh U, Prasad R, Kaur R. Agronomic interventions for micronutrient biofortification of pulses. Indian Journal of Agronomy. 2016;**61**(4th IAC Special Issue):161-172

[152] Maiti R, Rajkurmar D, Jagan M, Pramanik K, Vidasagar P. Effect of seed priming on seedling vigour and yield of tomato and chilli. International Journal of Bio-resource and Stress Management. 2013;**4**(2):119-125

[153] Dutta SK, Singh AR, Boopathi T, Singh SB, Singh MC, Malsawmzuali. Effects of priming on germination and seedling vigour of bird's eye chilli (*Capsicum frutescens* L.) seeds collected from eastern Himalayan region of India. The. Bioscan. 2015;**10**(1):279-284

[154] Manjunath BL, Dhanoji MM. Effect of seed hardening with chemicals on drought tolerance traits and yield in chickpea (*Cicer arietinum* L.). Journal of Agricultural Science. 2011;**3**(3):186-189

[155] Ashraf M, Rauf H. Inducing salt tolerance in maize (*Zea mays* L.) through seedpriming with chloride salts: Growth and ion transport at early growth stages. Acta Physiologiae Plantarum. 2001;**23**(4):407-414

[156] Anosheh HP, Sadeghi H, Emam Y. Chemical priming with urea and KNO_3 enhances maize hybrids (Zea mays L.) seed viability under abiotic stress. Journal of Crop Science and Biotechnology. 2011;**14**(4):289-295

[157] Arun MN, Shankara Hebbar S, Bhanuprakash K, Senthivel T. Seed priming improved irrigation water use efficiency, yield and yield components of summer cowpea under limited water conditions. Legume Research—An International Journal. 2017;**40**(5):864-871

[158] Harris D, Rashid A, Miraj G, Arif M, Yunas M. 'On-farm' seed priming with zinc in chickpea and wheat in Pakistan. Plant and Soil. 2008;**306**:3-10

[159] Zheng M, Tao Y, Hussain S, Jiang Q, Peng S, Huang J, et al. Priming in dry direct seeded rice: Consequences for emergence seedling growth and associated metabolic events under drought stress. Plant Growth Regulation. 2016;**78**(2):167-178. DOI: 10.1007/s10725-015-0083-5

[160] Devi K, Barua PK, Barua M. Integrated effect of pre-sowing seed treatment, sowing windows and seasons on seed yield and quality of green gram. Legume Research. 2021;**44**(8):956-961

[161] Shariatmadari Z, Riahi H, Seyed Hashtroudi M, Ghassempour A, Aghashariatmadary Z. Plant growth promoting cyanobacteria and their distribution in terrestrial habitats of Iran. Soil Science and Plant Nutrition. 2013;**59**:535-547

[162] Sarma BK, Yadav SK, Singh S, Singh HB. Microbial consortium mediated plant defense against phytopathogens: Readdressing for enhancing efficacy. Soil Biology and Biochemistry. 2015;**87**:25-33

[163] Lopez LVP, Rodríguez AR, Coronado MES, Hernandez PEM, Segovia AO. Effects of hydro priming treatments on the invigoration of aged *Dodonaea viscosa* seeds and water-holding polymer on the improvement of seedling growth in a lava field. Restoration Ecology. 2016;**24**(1):61-70

[164] Desmukh AJ, Jaiman RS, Bambharolia RP, Vijay AP. Seed biopriming—A review. International Journal of Economic Plants. 2020;**7**(1):38-43

[165] Pill WG, Collins CM, Goldberger B, Gregory N. Responses of nonprime or primedseeds of "Marketmore 76" cucumber (*Cucumis sativus* L.) slurry coated with Trichoderma species to planting in growth media infested with *Pythium aphanidermatum*. Scientia Horticulturae. 2009;**121**:54-62

[166] Reddy PP. Bio-priming of seeds. In: Recent Advances in Crop Protection. New Delhi: Springer; 2012. p. 83

[167] Kumar VV. Plant growth-promoting microorganisms: Interaction with plants and soil. In: Hakeem KR, Akhtar MS, Abdullah SNA, editors. Plant, Soil and Microbes:Volume 1: Implications in Crop Science. Springer, Cham; 2016. pp. 1-16

[168] Pehlivan N, Yesilyurt AM, Durmus N, Karaoglu SA. *Trichodermalixii* ID11Dseed biopriming mitigates dose dependent salt toxicity in maize. Acta Physiologiae Plantarum. 2017;**39**:79. DOI: 10.1007/s11738-017-2375-z

[169] Ilyas S, Asie KV, Sutariati GAK. Bio matriconditioning or biopriming with biofungicides or biological agents applied on hot pepper (*Capsicum annuum* L.) seeds reduced seed borne

Colletotrichum capsici and increased seed quality and yield. Acta Horticulturae. 2015;**1105**:89-96

[170] Vishwas S, Chaurasia AK, Bara BM, Debnath A, Parihar NN, Brunda K, et al. Effect of priming on germination and seedling establishment of chickpea (*Cicer arietinum* L.) seeds. Journal of Pharmacognosy and Phytochemistry. 2017;**6**(4):72-74

[171] Prachi S, Jyoti S, Shatrupa R, Rahul SR, Anukool V, Rakesh KS, et al. Seed biopriming with antagonistic microbes and ascorbic acid and induce resistance in tomato against *Fusarium* sp wilt. Microbiological Research. 2020;**237**:1-13

[172] Abadeh MR, Sharifi RS, Imani A. Influence of nitrogen and seed biopriming with plant growth promoting rhizobacter (PGPR) on yield and agronomic characteristics of red lenthil. Journal of Applied Environmental and Biological Sciences. 2013;**3**(11):117-123

[173] Kasim WA, Osman ME, Omar MN, Abd El-Daim IA, Bejai S, Meijer J. Control of drought stress in wheat using plant-growth-promoting bacteria. Journal of Plant Growth Regulation. 2013;**32**:122-130

[174] Shah G, Jan M, Afrien M, Anees M, Rehman S, Daud MK, et al. Halophillic bacteria mediated phyto remediation of salt-affected soils cultivated with rice. Journal of Geochemical exploration. 2017;**174**:59-65

[175] Sen SK, Mandal P. Solid matrix priming with chitosan enhances seed germination and seedling vigouration in mung bean under salinity stress. Journal of Central European Agriculture. 2016;**17**(3):749-762